AF477025

THOSE FABULOUS FLYING YEARS

Joy-Riding and Flying Circuses Between The Wars

Colin Cruddas

An Air-Britain Publication

Published in the United Kingdom by:
Air-Britain (Historians) Ltd.
12 Lonsdale Gardens,
Tunbridge Wells,
Kent, TN1 1PA

Sales Dept:
41 Penshurst Road, Leigh,
Tonbridge, Kent TN11 8HL

Edited by Malcolm P Fillmore

All correspondence regarding this book to:
Malcolm P Fillmore,
4 The Courtyard, East Park,
Crawley RH10 6AG
and not to the Tunbridge Wells or Tonbridge addresses

ISBN 0 85130 334 X

Printed in the United Kingdom by:
The Cromwell Press Ltd,
Aintree Avenue,
White Horse Business Park,
Trowbridge,
Wiltshire BA14 0XB

Origination by Howard Marks, Hastings

Front cover:
>   *Artist Colin Ashford captures a scene at a typical
>   Sir Alan Cobham's National Aviation Day in 1932 with
>   the all-red Cornwall Aviation Company Avro 504K,
>   a Cierva C.19 Autogiro and the Airspeed Ferry
>   waiting for another load of joy-riders.*

Contents

Dedication

*The preparation of this book has required me to draw on the knowledge, expertise and patience of many people.
Some have travelled with me throughout the whole project, and none more so than my wife Thelma.
Her consistent encouragement, constructive advice and tolerance (when essential gardening duties
often took second place to my scribing) have long been appreciated as crucial pillars of support.
Accordingly I dedicate this work to her with much love and grateful thanks.*

Thelma Cruddas about to depart from Old Sarum, Wilts in DH.60G Moth G-AAWO. The pilot/owner of this surviving joy-riding machine is the author's former Flight Refuelling Ltd work colleague, John Reid.

Sir Alan Cobham – 'Ringmaster Extraordinaire'.

Foreword
by Sir Michael Cobham CBE, MA, Barrister, FRAeS

Colin Cruddas' book deals with a fascinating period in the history of aviation namely the 1920s and '30s which were, and should be regarded as, still part of the very early years of the development of the aeroplane. It must be remembered that the famous first powered flight of the Wright brothers on 17th December 1903, although of immense importance, lasted but 12 seconds and covered a distance of only 40 metres, and it was not until one or two years later that flights of any serious dimension or duration took place – albeit that this inspired pair of pioneers were and remained leaders in the development of the aeroplane, certainly in the early years leading up to the First World War. That dreadful conflict did of course accelerate enormously technological progress and the development of aviation.

It is to me nevertheless remarkable that in less than twenty years from that first flight, flying demonstrations, mainly as an entertainment, were being undertaken on such a scale as took place in the '20s and early '30s. These activities are most meticulously recorded in this book, and are I believe of immense historical significance in the story of the development, and people's acceptance of the aeroplane, first as a growing and latterly as a major means of mass-transportation.

However, any book on 'barnstorming', as such shows came to be called, would have to refer to the contribution of Alan Cobham in the early '30s, for in one important respect he differed from the many others who were involved in these activities. This difference was that, whilst not overlooking the business aspect, Alan Cobham believed he had a mission – namely to bring aviation to the people and to use his favourite expression to spread "air-mindedness" especially to the younger generation. His endeavours in this respect were I believe rewarding not only to himself and to the young people he influenced, but also to the nation as a whole.

In this connection it is noteworthy that in the UK volunteers for aircrew duties in the Second World War when asked if they had ever flown, nine times out of ten responded if they had that their first flight was at an Alan Cobham Air Display. I personally have met many Second World War aircrew (several of whom later became senior officers) who told me that their first flight was with my father's Air Displays, or air circuses – a term incidentally which he intensely disliked, but which was nevertheless popularly adopted!

One aspect of these early years of aviation is I feel all too easily overlooked – namely its remarkably short timespan. Between 1903 and 1905 the first powered flights by the Wright brothers took place and yet, only twenty years later, Alan Cobham had flown from London to India, London to Cape Town and London to Australia and back. As a simple comparison, this timespan is the same as that which now distances us from the Falklands War, an event which to many of us is all too recent. Within but a further decade a sizeable network of air routes had been developed in the United States, and to a great extent between the capital cities of Europe, whilst the United Kingdom was also forging ahead with the routes to India and Africa.

These first thirty or so years were certainly fabulous in many respects, and historically were as momentous I believe as any in the whole history both of the development of the aeroplane and of air transport in all of its manifestations. This book which captures so well the atmosphere of the times makes compelling reading and I commend it to you.

Acknowledgements

To my knowledge, no previously published work has attempted collectively to describe the disparate aerial display and joy-riding activities that took place throughout the United Kingdom during the inter-war period.

Faced with this discovery, something, I concluded, needed to be done. Accordingly, in an attempt to track down the more elusive and all too frequently 'fly-by-night' operators, my research has thrown a net of enquiries over the entire British Isles. Having stirred the mists surrounding many hazy memories I am deeply grateful to the following persons for their patience and contributions to this story, some of whom, sadly but inevitably, having now left the stage:

Barry Abraham
Ken Aitken
Colin Ashford
Percy Allison (deceased)
Christopher Balfour
Frances Barton
Captain Cecil Bebb (deceased)
Peter Berry
Jack Brackenbury
Terry Brien
Robin Brooks
Russell Brown
Colin Bruce (deceased)
Martin Burney
David Collyer
Peter Connor
Air Vice Marshal Ron Dick
Roy Farquharson-Eskell
Nick Forder
Paul Francis (English Heritage)
Harry Frazer-Mitchell
Olive French
Brian Gardner
Steve Gillard (BAE Systems Heritage Centre, Brough)
Alistair Goodrum
Mike Grant
Peter Green
Malcolm Hall
John Hamlin
Ken Harman
John Havers
Harry Holmes
Mike Hooks
Sqn Ldr Tony Iveson
Captain P W Kennedy
John King
Air Chief Marshal Sir Michael Knight
Hayden Lawford
Stuart Leslie
Ted Lightbrown
John Maynard
David Mooney
Harry Morris (deceased)

Phil Munson
Mick Oakey
Tony Pearce
Grant Peerless
Mike Phipp
Frank Poller
Sqn Ldr Colin Pomeroy
Maureen Rawlings
Brian Riddle (Royal Aeronautical Society)
Richard Riding
Alan Rowley
Tom Samson
Graham Simons
Peter Skinner
David Smith
Graham Smith
Harry Smith
Dr. Hugh Thomas
Raymond Towler
Julian Temple
Harry Ward (deceased)
Guy Warner
Ray Wheeler
Ray Williams

My especial thanks go to author Ted Chapman and Captain David Phillips for permission to draw extensively upon the excellent 'Cornwall Aviation Company' and other family material, to Wg Cdr Malcolm Ward for access to his father Harry Ward's papers and to Alison and Andrew McGahey for the extended loan of the Holmes' photograph album. Phil Jarrett and Dick Riding have also been towers of strength in providing many specialist photographs.

In the 'highest awards' category I must pay special tribute to my wife Thelma and to Ernest Clear Hill for their critical appraisals of text and presentation and to Daryl Carmichael and Wendy Goodwin for their infinite patience and willingness to turn endless reams of barely decipherable text into a readable form. Sadly, Daryl's death following protracted illness, precluded her seeing the finished product, but her ever willing efforts clearly demonstrated a 'barnstorming spirit' that matches the courage of many described in this book.

My sincere thanks also go to 'Air-Britain's' chairman Malcolm Fillmore for his expert advice, editorial guidance and the unrestricted access to his invaluable database.

Finally I wish to express my deep appreciation not only to Cobham plc for the use of its archive, but also and especially to Sir Michael Cobham for his encouragement in what, at the outset we both considered to be an essential project.

Colin Cruddas
Shaftesbury, June 2003

Introduction

'Bathe was by Bladud to perfection brought,
By Necromanticke Arts, to fly he sought,
As from a Towre, he sought to scale the sky,
He brake his necke, because he soar'd too high'

Memorial of English Monarchs
John Taylor 1622

Following the world's first international aviation meetings held in France and Britain in 1909 and America in 1910, exhibition flying, air racing and record attempts became immensely popular on both sides of the Atlantic.

However, the aerial events that took place subsequently in Europe and the US essentially featured the same small elite group of fliers and were usually held close to large cities. They made little or no impact on the large proportion of the population which, at that time, lived in widespread rural communities. In such areas, horses rather than horsepower would continue to dominate the local scene for at least a further two decades.

Even though many thousands of aircraft were produced and training facilities were greatly expanded during the First World War, the aerial activity created by this prodigious effort occurred mainly over Europe's battlefields. The vast majority of British citizens remained largely unaware of the advances which had taken place in aircraft design and manufacture and also of the general improvements in piloting skills. But, by 1919, following the restoration of peace, a new and daring development began to emerge which would truly sow the seeds of 'bringing aviation to the people'.

In Britain alone, the number of testosterone-fuelled young ex-Royal Air Force pilots facing the bitter chill of unemployment was over 22,000 and this was matched by nearly 9,000 Americans who had received wartime flight training with the US Army's Air Service. Though, naturally, not all wanted to continue flying, many hoped to put their newly-found, indeed often sole capability, to good purpose and took the opportunity to purchase the relatively cheap training and fighting machines that, like themselves, had now become surplus to peacetime requirements. The age of the free-roaming aerial exhibition artiste willing to provide joy-rides to anyone brave or foolhardy enough to challenge the 'third dimension of the air' had arrived.

In America, the evocative term 'barnstormer', originally used to describe theatrical touring groups that usually entertained in the biggest local barn, was now extended to include this new, racy, roistering breed of airborne individuals. Impressively attired in military style costume and sporting impressive moustaches, they criss-crossed the country willing to risk not only their own lives, but also those of their customers, in the constant search for a few bucks. Times were hard however for these gypsy fliers with their deeply etched tans. Frequently it required all the verbal skills of the fast-talking confidence trickster to separate a reluctant public from its hard-earned cash.

All too often a pilot would arrive at some out-of-the way destination, with a last gasp of fuel in his tank, to find that any disposable dollars or willingness to venture aloft had been thoroughly drained but a few days previously by some other wandering 'skywayman' who, like himself, was just passing through. More often than not, the early 'barnstormers', concerned with simply surviving from one day to the next, found that any money raised was immediately absorbed in providing the next meal, a much needed bath or a gallon of petrol and patching up a barely airworthy machine.

As the desperate days of the early 1920s slipped by, the need grew for the itinerant flier to extend his repertoire of aerial stunts and manoeuvres. But the price tag that came with the increased risk-taking was high and a wide trail of crashed aircraft and broken bodies soon bore testimony to a game that had got out of hand.

Whilst in Britain and many other countries the control and regulation of civil aviation had been introduced immediately after the War, the position in the USA may be seen as having remained in the 'Wild West'! It was only in 1926 that the U.S. Federal Government, concerned at the level of accidents related to aviation – 179 in 1923 alone which resulted in eighty-five deaths and 126 serious injuries – introduced the Air Commerce Act. Somewhat akin to Britain's Air Navigation Acts of 1911 and 1920, this required, for the first time, official registration of aircraft and licensing of pilots and this brought about the rapid demise of the lone maverick performer. But, it also encouraged the formation of what, in many cases, became professionally organised touring groups.

Although the rural conditions that encouraged the arrival of the 'barnstormer' (a term incidentally rarely used at the time in Britain) were broadly the same both in America and England, it was the sheer flamboyance and the extremes of lunacy introduced by the less restrained American fliers that in some cases, influenced and inspired their British counterparts.

Many volumes have been produced describing in great detail the exploits of the American performers; nevertheless, an overview of their activities is included here to set the scene against which the early British pioneering efforts were made to make the 'man in the street' airminded.

Although, as will be seen, many individuals and organisations played significant parts in the widespread joy-riding phenomenon that overtook Britain in the 1920s and 1930s, the unique contributions made by Sir Alan Cobham, the Holmes brothers and their contemporary associate Captain Percival Phillips can perhaps be said to best epitomise the story of British 'barnstorming'. Though largely concentrating on their exploits, this book will, I hope, provide a fitting tribute to all the men and women entertainers who demonstrated their craft so bravely in the aerial arena.

A Curtiss Jenny awaiting the next brave customer to 'dice with death for a dollar'. This appears to be a still from the 1957 film "The Spirit of St.Louis" about the life of Charles A Lindbergh starring James Stewart.

An athletic performer demonstrates a 'plane to auto' transfer at Atlantic City in 1921.

Before his 1927 Atlantic flight, Charles Lindbergh was a well-known barnstormer, billed as 'Daredevil Lindbergh'.

Chapter One: Early Days – Over There

*'I don't do those things because I want to run the risk of being killed.
I do it to demonstrate what can be done. Somebody has got to show the way.
I want to do things that people feel can't be done. I don't believe anything is impossible
but perpetual motion. I am convinced that someday we will all be flying and the more
things that are attempted and accomplished, the quicker we will get there.'*

Ormer Locklear, 1919

'Exhibition' flying has been around in one form or another for as long as aviation itself. Since the earliest aeronauts made their first tentative ascents in balloons, gliders and later, flimsy 'stick and string' machines, challenging questions concerning how far, how high and, more specifically, how fast would the latest aerial carriage transport itself, have always been asked. Not surprisingly there has never been any unwillingness on the part of the designers, pilots or the owners of machines to show-off their charges in the best and most competitive light.

Today, a new aircraft's ability to meet a customer's basic performance requirements is almost taken for granted. It is the searching demands for optimum economic operation, environmental compatibility and high-level reliability that are now more likely to tax a manufacturer's ingenuity. Such latter-day considerations as these, which now are crucial elements in the 'total cost of ownership' equation, were simply of no account when the new science of aeronautics really began to flex its wings in the early part of the Twentieth Century.

In those far-off days, it still cost a great deal of money, time and effort to build a machine which, in the general absence of published design data, relied more on intuitive know-how for even a fighting chance of success. Added to this, the lack of dependable, lightweight engines created formidable problems for the pioneer constructors. The progress of aviation was therefore dictated by a dedicated and highly motivated elite group of questing pioneers who guarded their hard won knowledge closely.

None more so than the Wright Brothers, whose understandable reticence regarding the public documenting of trials and research work that surrounded the successful first powered flight programme amounted almost to an obsession. But it was whilst maintaining this isolationist, albeit justified and business like attitude, that the public spotlight in America swung onto Glenn Curtiss, a flamboyant motor-cycle manufacturer, then based in Hammondsport, New York.

Curtiss, whose range of light-weight engines was eagerly sought by other early constructors, was recruited in 1907 by Dr Alexander Graham Bell to join his newly formed Aerial Experiment Association. This group's self-appointed task was to build an aircraft that would rival the Wright Flyers whose performance, whilst still shrouded in secrecy, was reported through leaked information to be rapidly improving.

Glenn H Curtiss, who combined engineering skills with an outstanding business talent to challenge the Wright brothers.

This aim was achieved when, in the summer of 1908, Curtiss demonstrated the 'June Bug' for the first time. The Wrights, having discovered that this aircraft blatantly incorporated control features supposedly protected by their patents, were then forced into adopting a more competitive stance and to promoting keener awareness of public interest in their work.

At no time however, did the Wrights' resolve to protect their invention from unscrupulous 'copy-cat' designers weaken, nor their determination to acquire great wealth from their genius falter. In 1910, Curtiss and the Wrights were deeply engaged in a bitter legal battle over patent infringement and were sparing no effort to outdo each other with their respective Exhibition companies.

The Moisant International Aviators was another sensational touring outfit that, with perhaps a hundred or so other fortune seeking aerial performers, now also provided unwelcome competition for Curtiss and the Wrights.

From 1909, large air gatherings, along with national air races and record attempts really took hold of the public's imagination on both sides of the Atlantic. In America, closely related exhibition and stunt flying events were guaranteed to bring in the crowds to witness the mainly friendly rivalries and frequent crashes that provided extra excitement and uncertainty to the aviators' performances. Typical of these were the classic pre-war encounters between prominent pilots of the day and big name racing car drivers such as 'Crazy Eddie' Rickenbacker and Barney Oldfield. The close proximity of spectators to the action ensured that they would frequently be engulfed in clouds of dust as the pilot roared in at ground level behind the driver. He, fighting desperately to keep his powerful car on a narrow rough-hewn track, would be subject to every trick in the pilot's book as he tried to force him off it.

Behind the general camaraderie, however, there were occasions when less well-intentioned competitiveness gained the upper hand. An example of this occurred in 1910, when the French aviator Louis Paulhan was prevented from competing in prize events in San Francisco. Having uncrated his machine, he found that it had been severely damaged by saboteurs and he had to return home, furious and well out of pocket.

By now, many first class fliers of both sexes, were beginning to make their mark on the American flying scene. Fame, however, came at a high cost with, in many cases, death the inevitable outcome. Faulty design calculations, poor choice of materials or substandard assembly methods often resulted in frail machines experiencing catastrophic mid-air structural failure. Such accidents were perhaps unavoidable at this rudimentary stage of design and manufacturing development. Nevertheless, all too often, inadequate maintenance of airframes and engines or simple plain carelessness on the part of the pilot contributed to the large number of fatalities.

One early victim of seeming over-confidence was Lincoln Beachey who had flown for Curtiss, was the first American to loop and fly inverted and was regarded by Orville Wright as 'the greatest aviator of all'. On 14th March 1915, he was requested by officials at the San Francisco Panama Pacific Exposition to perform in his new and still largely untried Special Monoplane. Beachey agreed to carry out his sensational 'Death Dip', a manoeuvre he had successfully undertaken on hundreds of previous occasions in his sturdier Little Looper biplane. But this time the monoplane's wings folded and he plunged into the Bay before 50,000 horrified spectators. Pertinently and dramatically he had once written: "The Silent Reaper of Souls and I shook hands. Thousands of times we have engaged in a race among the clouds – plunging headlong in breathless flight – diving and circling with awful speed through ethereal space. And, many times, when the dazzling sunlight has blinded my eyes and sudden darkness has numbed my senses, I have imagined him close at my heels. On such occasions I have defied him, but in so doing experienced fright which I cannot explain. Today the old fellow and I are pals."

Another exponent of this era was the brave and beautiful Harriet Quimby. Invariably dressed in an eye-catching purple flying-suit, she became the first American woman to gain a Pilot's Certificate. Normally the most cautious of flyers and widely admired as the first lady to fly across the English Channel on 16th April 1912, she was lost, tragically, less than three months later. On 1st July 1912 whilst flying a Bleriot with a passenger, William A P Willard, her aircraft hit a sudden gust and both occupants, neither wearing any kind of restraint, were pitched out of the machine into Boston harbour.

Some of the early exhibition aviators saw active service in Mexico during the pre-war period. One such adventurer was Hector Worden, who, in 1911, was hired by the Mexican Constitutional Government to experiment with bombing and scouting techniques. Despite the highly attractive financial inducement of $1,500 per month, the frequency with which bullets raked his plane soon caused him to revert to his former supposedly less high-risk occupation of air display pilot. Fate, however, played the final card for, flying at a state fair in Dallas, Texas, soon after his return, he had a heart attack and according to medical evidence, died before he struck the ground.

America's entry into the European inferno in 1917 engendered a large aircraft manufacturing industry and the training of pilots on a commensurate scale.

Whilst there was a wartime ban on civilian flying, by 1919, flying schools had again begun to flourish, especially in California. Here, for example Cecil B De Mille, who was to become no mean aviator himself, was one of the first film moguls to introduce aerial stunts into motion pictures. Such was his interest that he purchased and operated two flying fields near Hollywood and established the Mercury Aviation Company to house a stable of

Lincoln Beachey, whose daring earned him the title 'The Man who Owned the Sky'.

Harriet Quimby, who learned to fly at the Moisant Aviation School, was a highly skilled pilot on the American exhibition circuit.

aircraft and daring pilots possessing the extra skills required for stunt flying.

The end of hostilities resulted in many newly discharged pilots seeking work where they could. Some invested their hard earned demobilisation dollars in war surplus training machines, such as the Lincoln Standard J.I, or, more usually, the ubiquitous Curtiss JN-4D.

Curtiss's machine, which undoubtedly had proved to be 'the right aeroplane at the right time' for the US Army's training needs, came about following the merging of two separate design talents. During a visit to England in 1912, Glenn Curtiss, by then America's leading designer and manufacturer of aircraft met B Douglas Thomas, a twenty-two year old designer working with the Sopwith Company at Kingston-upon-Thames.

Both men held a common belief that the 'tractor' method of propulsion, ie with the propeller mounted in front of the aircraft, represented the next major development step forward. At that time 'pusher' type layouts with the engine and propeller located behind the pilot were still generally favoured. An agreement was reached whereby each man would produce independent designs which

Curtiss would then build. Thomas mailed drawings of his Model 'J' from England and Curtiss correspondingly drew up plans for what became his Model 'N'. Good as his word, Curtiss then assembled and test flew both types. The best characteristic of each was combined, resulting in the JN series.

The 'Jenny' – capable of 75mph flat out and fully aerobatic – had arrived. Now, as the favourite mount of America's post-war aviators it took to the air over California, the Mecca where the scent of orange groves rose into a seemingly endless blue sky, and where, more importantly, flying for the movies promised big rewards!

Not all pilots, however, possessed the ability to make the transition from service to civilian flying. Fewer still could offer the flying talent demanded by Hollywood for promoting the 'Ace Culture' in movies such as Howard Hughes' epic war film *"Hell's Angels"*. For many, the spectre of unemployment remained forever round the next corner, but for the select handful prepared to do whatever the producers required, fame and fortune beckoned.

Today the names of the early movie pilots are largely forgotten but when the 'Roaring Twenties' began, they were hailed as the cream of the daredevil crop. Restless almost to a man, their flying careers, punctuated by continual crashes – both intentional and otherwise – became colourful freewheeling chases to where the next large pay cheque was thought to be. Only short-term contracts were ever offered by the major studios and the intense competition within the stunting fraternity meant that an individual's earnings, whilst sometimes high, were more often, for long periods, virtually non-existent.

Consequently, of necessity these solo artists also undertook a mixture of exhibition, stunt, instructional and charter work. The latter frequently involved the carriage of individuals, guns or drugs across the ever-troublesome Mexican border. And, added to this of course was 'barnstorming'. This was real 'barnstorming', which included not only the piloting of the aircraft, but, in many cases 'wing-walking', aerial acrobatics and parachute descents, then regarded as the ultimate high-risk occupation. As one old-timer famously commented, "Dangerous it may have been, but it sure beat starvin' to death."

It was on 12th November 1921 over Long Beach, California, that stuntman Wesley May, standing on the top wing of a Lincoln Standard J.1 biplane with a five-gallon can of fuel strapped to his back (and no parachute) caught the wing-tip skid of a 'Jenny' flying just above and hauled himself aboard. His hazardous venture is now popularly acknowledged to be the first recorded instance of in-flight refuelling although it has to be said that most of the fuel was swept away in the slipstream. Wesley May sadly came to a bizarre end when, soon after, a parachute descent went disastrously wrong. Arriving with apparent safety in a tree in a graveyard, he released his parachute and fell onto a tombstone fatally fracturing his skull.

The ladies too participated in 'barnstorming' in no uncertain manner, bringing glamour and plain old-fashioned courage to the scene, matching and complementing the zany flying skills of their adventurous male colleagues. The public flocked in huge numbers to see slim, agile teenagers such as Gladys Roy walk blindfold from one wingtip to another, or Margie Hobbs, billed as 'Ethel Dare – The Flying Witch', change planes in mid-air, which she did ninety-five times before superstitious pilots concerned for her safety refused to take her up any more. Lillian Boyer, required little persuasion to give up her job as a waitress in Chicago to suspend herself by or from the underside of her aircraft. During a remarkable eig accident-free career, she commanded a daily fee that exceeded the amounts most 'barnstormers' earned in a whole year!

On 15th May 1921, Laura Bromwell set a women's record by completing 199 'loop-the-loops' over Garden City, New York before running low on fuel. Sad to relate however that, like Harriet Quimby, she failed to take adequate restraining precautions, when less than a month later, on 5th June she was pitched to her death whilst trying to loop an unfamiliar aircraft.

Looping had become the latest craze to capture the public's attention and aerial acrobat Laura Ingalls was paid a dollar a loop at exhibitions. Eventually she put Miss Bromwell's record in the shade, by completing 980 consecutive 'up and over' manoeuvres – a feat which stood admirable comparison with 'Loop-the-Loop' Murphy's looping all the way from San Francisco to Los Angeles!

Consider too the achievements of Bessie Coleman *[right]* who, born in Texas in 1893, was forced by racial prejudice to leave America for France in order to learn to fly. After becoming, on 15th June 1921, the first woman to earn an International Aviation Licence, she returned to provide aerial performances that ranked with the best. Affectionately known as 'Queen Bess', she became the country's first coloured aviatrix as she 'barnstormed' her way across the US before she too, on 30th April 1926 and aged thirty-three, fell to her death at Jacksonville, Florida. Her fortitude  however overcame the then widely-held perception that non-whites inherently lacked an aptitude for flying. Her hard won success paved the way for the first all-black air show, the Coloured Air Circus, held before a crowd of 15,000 on Labor Day, 1931.

Eddie Stinson displays his prowess as an aerial acrobat below a Curtiss JN-4; a stunt regularly performed in the 1920s.

An early form of in-flight refuelling, Wesley May undertakes a 'plane change with a five-gallon can of petrol strapped to his back' – this stunt took place over Long Beach, California on 12th November 1921.

These are but a few of the illustrious band of heroes and heroines who thrilled millions before officialdom and common sense eventually placed a restraining hand on the aviation show world.

The quest for new tricks continued to prove a constant challenge, whether it was for the thrilling sequences performed for the some 250 aviation-related films produced in the inter-war period, or for those carried out at state fairs and other public events. Eventually it could be claimed that every conceivable flying combination of man and machine had been demonstrated – or at least attempted. The transfer of acrobatic performers from plane-to-plane, in the case of a man called Bonnie Rowe with one hand tied behind his back, plane-to-car, train, boat or the reverse, all became routine acts. Aerial manoeuvres took place in which both men and women walked all over the outside of the aircraft without parachutes. All the time they battled against a raging slipstream and the sudden gut-wrenching sensation as their precarious perch was threatened by a violent downward gust. Such activities would stupefy the imagination of today's air show spectators, let alone the organising officials!

As William H Pickens, the shrewd and compassionless 'barnstorming' promoter, reminded a dazed 'wing-walker', Ormer Locklear, after one automobile-to-aeroplane change had gone seriously wrong, "Don't forget that we're both capitalising on sudden death. Bandages are box office and they hold the romance of freshly healed accidents and the lure of catastrophes to come". Sadly, this philosophy became grim reality when Locklear, the stuntman whose star briefly shone brighter than all others, died along with his pilot 'Skeets' Elliott in a spectacular Hollywood crash on the evening of 2nd August 1920.

It is thought that in carrying out a night flight for the film *"The Skywayman"* which required Locklear to stand on the wing, Elliott became disorientated by searchlight beams or perhaps the glare from pyrotechnics attached to the aircraft. His fiery end certainly accelerated the distribution of the movie in which Locklear posthumously gained star billing. The funeral arrangements were impressive even by Hollywood's excessive standards, with a full military guard of honour, a police band, a mounted posse of movie cowboys and aircraft circling overhead.

Pickens did not attend the funeral. "After all," he later claimed, "the show had to go on" and the task of finding a replacement aerial wonder man was clearly of greater importance than the demise of his erstwhile star. Indeed, just a week later, Barr's Illuminated Aerial Circus appearing at the Pickering Pleasure Park in Santa Monica, made a special point of advertising the same stunt that had brought about the deaths of Locklear and Elliott as a nightly feature.

But, in parallel with such spectacular performances, went the grinding quest by many other less noted aviators to persuade the nation's country folk to fly. As one veteran pilot, 'Slats' Rodgers, later reflected – "the lone fliers were a mixture of the Old West cowhand and the hot-rod driver of today. Once in the air they considered themselves free as the birds, just as the cowhand thought he was free as a coyote. Suckers were 'taken' wherever they could be found which meant fliers roaming the face of the earth like gypsies."

Perhaps the most famous aviator to cut his teeth on the country 'barnstorming' circuit was Charles A Lindbergh. In 1922, aged twenty and then living in Lincoln, Nebraska, he had already decided on an aeronautical career instead of one in politics like his father. Considering the uncertainties then attached to flying it hardly seemed a wise choice. Eventually he won parental approval and, billed as 'Daredevil Lindbergh – Wingwalker and Parachutist', he toured the mid-west, northern and southern states for a couple of seasons.

In his own account of those travels, Lindbergh recalled that whilst in Mississippi, a coloured lady of ample proportions approached him to enquire how much he would charge to fly her to Heaven and leave her there. Unfortunately his reply is not recorded. After a few years' subsequent service in the US Army Air

The Gates Flying Circus was perhaps the most successful US touring display in the 1920s.

Service, Lindbergh went on to become the hero of an entire generation when in May 1927, he undertook the first solo crossing of the Atlantic. The rest, as they say, is history!

The introduction of the Air Commerce Act in 1926 threatened the 'barnstormers' freewheeling way of life and sealed forever the fate of the itinerant and all too often highly irresponsible lone flier. The Act required aircraft to be registered and proven airworthy. Pilots had to be licensed and formally proficient, and aerial performers had to wear parachutes. This was something no self-respecting 'wing-walker' or plane-change artist had ever considered remotely acceptable. These mandatory changes proved draconian and too expensive for most of the single operators and the embryonic circuses which were still trying to squeeze a living out of a populace that, by then, could reasonably claim to have 'seen it all before'.

Many operators saw the new legal requirements as devastating to their livelihood but the far-sighted ones realised that the professional standards being introduced simply had to be adopted if aviation was to prosper. Accordingly several entrepreneurs who ran the more successful shows accepted the inevitable changes and even went on to advertise their revised attractions as bigger, better and more spectacular than before.

One such promoter, Ivan R Gates, and his partner Clyde 'Upside Down' Pangborn, took the Gates Flying Circus to more than 500 cities between 1919 and 1931 (by which time it had been renamed The Flying Fleet). Perhaps no-one epitomised the mad-cap jazz age better than the extrovert, flamboyant Gates. Although it was Pangborn, unassuming, modest and sincere, who bore the responsibility for the flying side of the business, it was the unstable, unreliable and on many occasions violent showman, Gates, who somehow steered the outfit through a decade of successful operations, entertaining an estimated fifty million citizens along the way.

Entertained the crowd certainly were – on one occasion, Rosalie Gordon, Gates's star 'jumper', found herself dangling helplessly some twenty feet below the aircraft, with the lines of her parachute caught up in the undercarriage. 'Wing-walker' Milton Girton climbed down from the cockpit and tried without success to haul the girl up. Even when a second stuntman, Freddy Lund, was transferred from a formating aeroplane their combined efforts were still insufficient to reel the girl in. It was only when the sturdier Pangborn handed over piloting the aircraft to the slightly built Lund that enough muscle power became available to pull her, inch by inch, up to the comparative safety of the undercarriage spreader bar. Pangborn then returned to the cockpit and landed the aircraft with great care before an audience convinced that it was all part of the normal performance!

The prosaically named Gates Flying Circus became one of the most enduring of the American flying troupes. Many others however enjoyed success over shorter periods. The Great American Flying Circus, 13 Black Cats, Love Field Lunatics, Doug Davis Flying Circus, Mabel Cody Flying Circus, Flying Aces Air Circus, the Fordon-Brown National Air Show, Dean's Flying Circus which featured Marie Mallaire, the 'Vamp of Cloudland', Dale Seitz's Safe and Sane Flyers and Pancho Barnes' Mystery Air Circus were just a few of the more exotically named 'squadrons' which took to the skies, each endeavouring to outpace the rest.

One of the most awe inspiring and hair-raising manoeuvres that took place were those performed by Ben F Gregory and Harold S Johnson's big corrugated metal-skinned Ford 4-AT tri-motors. Gregory's *"Ship From Mars"* was regularly featured at mid-west state and county fairs when, in addition to providing joy-rides, it became the showpiece attraction in dramatic late night grand finales. After several low passes over a darkened field filled with a hushed and expectant crowd, a pyrotechnic display would suddenly burst into life simulating anti-aircraft fire as, according to the excited commentator, the 'intruder' aligned itself for an 'aerial attack'. Then, without warning, a battery of floodlights mounted within the aircraft would be directed at each window and the open door. Intermixed with the exploding fireworks, the Ford's evasive aerobatics apparently created the effect of a gyrating fireball that was by all accounts truly spell-binding.

Johnson's performances were guaranteed to bring thrills to any air show. Not only did he spin, snap-roll and stall his six-ton machine at less than 1000 ft but he always finished his act with three consecutive loops 'off the deck' with the last one terminating in a single wheel landing. What made this feat so amazing was that in conducting each loop, Johnson, single-handedly, had to adjust the tailplane trim from positive to negative by sixty-two turns of a hand crank while simultaneously manipulating three throttles and ensuring that he arrived at the bottom of the loop with, hopefully, twenty-five feet of height to spare!

American aviation's original rip-roaring image became overtaken by the introduction of safer, yet still exciting, displays that embodied a more responsible attitude towards the paying public. No longer did performers trade on the panache and glamour associated in the public mind with military wartime flying. At the insistence of publicity-conscious airlines, aircraft manufacturers and other vested interests, the introduction of the Civil Aeronautics Act in 1938 tightened the airshow regulations still further. This proved to be the last nail in the coffin of the old style flying circuses and the final demise of the itinerant passenger-seeking state-hopping opportunist pilots. By that time however, most of the really big names had flown into the final sunset. Few had survived to retire.

"The Ship from Mars" was Ben Gregory's Ford 4-AT. It provided a spectacular fiery finale to displays.

In October 1909, Blackpool [seen here] and Doncaster both claimed to have held the first aviation meeting in England.

Chapter Two: Meanwhile – Over Here

'Like soldiers, pilots never die, they only fade away,
And off they glide to better lands where life is always gay,
With lots of lovely aerodromes on each and either hand,
And underneath them all the time, huge fields in which to land.'
Piffling Poems for Pilots – Pauline Gower 1934

At the time of the Wrights' first engine-driven flight in 1903, aviation progress in Europe appeared to have lost its way. The Nineteenth Century had seen many innovative attempts by European man to fly and the considerable number of designs that never progressed beyond the drawing board is, perhaps, ample proof of the determination in the Victorian era to achieve powered flight. But the brave efforts of British, French, German and Russian pioneers produced relatively few practical results. Germany's Otto Lilienthal and the Englishman Percy Pilcher, who between them made hundreds of successful glider flights between 1891 and 1899, were, however, notable exceptions to the rule.

In 1902, Continental interest was rekindled when a Frenchman, Captain Ferdinand Ferber, produced copies of a Wright glider but whilst he and others, concentrated on developing successful non-powered machines, in America, in 1905 the Wright Flyer III was flying for thirty minutes and carrying out an impressive range of manoeuvres including fully banked circles and figures-of-eight.

In Europe, Alberto Santos-Dumont, the diminutive Brazilian whose balloon and airship flights were a familiar sight over Paris at the turn of the century, was commonly believed to have made the first breakthrough in powered flight when his aeroplane No.14 bis ('encore' or 14a) made a sixty-five metre hop on 23rd October 1906. It was not until the Wrights brought their Flyer to Europe in 1908 and demonstrated the ease with which it could out-perform any competition, that Santos Dumont eventually allowed the Americans' claim to have introduced powered flight.

Radical developments were, however, about to change the aviation scene. New names emerged in France such as Bleriot, Voisin, Farman, Breguet, Levavasseur and Robert Esnault Pelterie which, at first challenged, and then surpassed Santos-Dumont, to become Europe's leading designers. In Britain, aeronautical progress was also made when 'Colonel' Samuel F Cody made the country's first officially recognised powered flight (424 metres) at Farnborough on 16th October 1908. Within the year, on 13th July 1909, Alliott Verdon Roe also made headlines when he became the first British subject to fly an all-British machine, the Roe Triplane, at Lea Marshes in Essex. Less than two weeks later, on 25th July, in what proved to be a momentous year, Louis Bleriot made his epic flight across the English Channel.

Following the world's first aviation gathering at Rheims in northern France between 22nd-29th August, 1909, which featured thirty-eight aircraft, major British flying meetings were held at Doncaster, 15-25th October 1909 and at Blackpool, 18-29th October 1909, that each attracted mid-week crowds of over 50,000 people.

It is of interest to note that, whilst at these meetings aircraft all too often failed to leave the ground, by the following year, their ability to fly had much improved. Nevertheless, the uncertainties attached to becoming airborne were clearly recognised at Blackpool where, at flying meetings, flags were flown on top of the Tower to depict the possibility, even the probability, of aviators leaving the ground. A black flag indicated 'no flying at present', a white flag denoted that 'flights will take place shortly', and a red

Bentfield Charles Hucks, piloting his Bleriot XI, was a popular flyer at pre-war 'air meets'.

flag told the crowds that 'flying could take place at any time'. This final optimistic signal was guaranteed to send enthusiastic hordes rushing to the promenade for the trams that would take them to see the 'gods' perform.

It was also in 1909, in February, that Short Brothers, founded on the Isle of Sheppey by Horace, Eustace and Oswald, became the world's first aircraft manufacturing firm when it was awarded a contract worth £8,400 to construct six Short-Wright Flyer biplanes for the Aero Club of the United Kingdom (which became the Royal Aero Club in February 1910).

The aviation bug was also beginning to bite in other areas of the country. In Yorkshire, a marine engineer, Thomas Bell, made a highly commendable effort to win the substantial prize of £1,000 offered by the *"London Daily Mail"* for the first British aeroplane to fly a mile over a closed circuit. His Newington Monoplane, named after his backers, the Newington Steam Trawling Co, was placed on public view in the Hull Royal Institute before removal to nearby Hedon for flying trials. Unfortunately, the machine was written off after an over-zealous financial colleague attempted to fly it for the first time. The prize was eventually won by JTC Moore-Brabazon who flew the required distance at Leysdown, on 30th October 1909 in the Short No.2 biplane.

In that halcyon pre-war period, considerable public interest was shown in the flying displays that took place at more central venues near London, such as Hendon and Brooklands featuring such charismatic fliers as Gustav Hamel, Freddie Raynham and Bentfield Charles Hucks. But as far as Mr and Mrs Everyman were concerned, it was largely a question of "look, but don't touch"; with limited opportunities even for those who could afford it, to actually experience the thrill of becoming airborne themselves.

Two performers who were determined to get passengers into the air were Yorkshire's Roland William Ding and Harold Blackburn (unrelated to the aircraft constructor Robert Blackburn). In 1914, they flew in friendly competition carrying passengers from The Stray in Harrogate. Ding, then chief pilot for the Northern Aircraft Company, flew the distinctive crescent-winged Handley Page Type G/100 biplane whereas Blackburn piloted an early version of the Avro 504, the machine destined to become the mainstay of joy-riding operations in the post-war years to come.

Roland Ding and the unusual crescent-winged Handley Page G/100 at Holray Park, Carlton in June 1914; the aircraft crashed a month later landing at Northallerton carnival.

Although it was his French rival, Louis Paulhan, who won the honours in the 1910 *"Daily Mail"* London to Manchester air race, Claude Grahame-White gained the nation's praise and headlines when, in trying to overtake Paulhan, he became the first aviator to attempt a night flight in Great Britain. His celebrity then took him on the first of two trips to America where, taken to heart by the country's social elite, he commanded enormous fees to appear at the air meetings and racing events that had now become a major spectator sport.

Following his return to England in 1911, he moved his flying operations from Brooklands to Hendon and set about, in partnership with Louis Bleriot and Sir Hiram Maxim to create an international aviation centre. The company's ambitious prospectus called for a share capital of £200,000, but the issue failed when public subscriptions fell far short of this amount. Undeterred by this set-back, Grahame-White immediately set about forming the Grahame-White Aviation Co and acquired a ten-year lease on Hendon's 207 acres. He also obtained the right to purchase the freehold and additional land for what was now to become the London Aerodrome.

Shortly before the outbreak of war, and no doubt attracted by the prospect of appearing before crowds in his home county, Hampshire-born Grahame-White carried out a physically demanding tour of the south coast. This involved his visiting 121 towns, giving 500 exhibition flights and carrying 1200 passengers. Determined to instil aviation consciousness in the public's mind he emblazoned the sprightly slogan *'Wake up England'* on the fuselage and wings of his Farman machine. His words of exhortation, although aimed primarily at encouraging support for civil aviation, were particularly prescient considering the imminence of forthcoming hostilities. As a propaganda exercise, his ambitious programme was later regarded by many as comparable in importance to those undertaken in the post-war years by another arch-campaigner for 'air-mindedness', the legendary Alan Cobham.

After the end of hostilities in 1918, all the organisations linked to war product manufacture faced an uncertain future. Whilst, in America, a Curtiss JN-4D could be purchased for some $500 (then the equivalent to £125), less than a tenth of its original build price, the British equivalent, the Avro 504K, was made available by the Aircraft Disposal Board for about £500. But, this was a considerable sum in those days and far beyond the reach of most would-be aviators.

Alan Cobham, whose personal involvement was to play such a critical part in the evolution of British aviation, later stated that for the 22,000 trained pilots demobilised in 1919, less than one job per thousand existed in civilian aviation. Cobham, himself an ex-RAF flying instructor, was however more determined than most to continue his flying career.

His first peace-time appointment in early 1919 was with the Willesden-based British Aerial Transport Company Ltd which produced a series of biplane and sesquiplane machines (FK.22-FK.28) designed by Frederick Koolhoven. However, Cobham's efforts to find landing sites suitable for pleasure flights at East Coast resorts lasted for only one month before the company changed its plans.

Then, as the result of an advertisement placed in the Aerial Register and Gazette, he contacted two brothers, Frederick John Vernon (Fred) and John Duncan Vernon (Jack) Holmes, also newly discharged from military service. Pooling their limited savings in April 1919, they formed the Berkshire Aviation Company with the express aim of providing pleasure flights within the county. Yew Tree Farm (then owned by the White family but taken over soon afterwards by Richard Rickards) in the Holmes' home village of East Hannay, near Wantage, became their maintenance base and testing was also carried out at The Common, a field at nearby Lyford, which belonged to Richard's brother, Charles Rickards.

Claude Grahame-White sported this pertinent message when touring in 1912 in response to officialdom's indifference to aviation in England.

When civilian flying was allowed to resume in Britain on 1st May 1919, it is conservatively estimated that some fifty ex-Service fliers with similar thoughts to Cobham and armed with great enthusiasm but little experience, set out to entice the British public into the air at a guinea (105p) a head. This represented a fortune to most people and thus few of these small outfits survived more than one or two years of operation.

But the Berkshire Aviation team, with assets consisting of one Avro 504K, D9298 (later G-EACL), the first to be sold by the Disposal Board at the Waddon/Croydon depot, one Ford car, a stock of petrol and financial reserves amounting to £200, was perhaps the best organised to meet and overcome the various administrative difficulties. The most daunting of these was the Air Ministry's insistence on inspecting and (eventually) approving every field or landing ground intended for use. Clearly, when faced with the additional vagaries of weather, mechanical breakdown and frequent last minute changes of mind exercised by farmers and landowners, this unwieldy and time-consuming approval system caused a great deal of frustration.

Berkshire's initial plan was to visit eight different locations in its home area over a three-week period, but the hiring of suitable fields proved difficult. As Cobham recalled in his autobiography 'A Time to Fly', when all else failed it was left to an appeal to the farmer's better nature along the lines of - "We've just come out of the RAF and are trying to make a living. We did our bit then. Can't you help us now?" This invariably won the day!

When applying to the Air Ministry for licences for the first three flying fields, Cobham was told that no-one was immediately available to process the paperwork and that it would take several weeks for the approvals to arrive. Upon being informed that this was just the beginning of a new and exciting enterprise the less than enthusiastic official in charge, no doubt fearing an imminent increase in workload, threatened to forbid the whole operation. It needed very persuasive arguments indeed, pointing out that the very interests of civil aviation were at stake finally to overcome bureaucracy and to get a special office opened to deal with the extra work. It did not go unnoticed however that the 'heavily overworked' Air Ministry staff later earmarked for these duties, invariably turned up for work between ten and ten-thirty each morning and enjoyed half-hour tea breaks either side of an extended lunch period before leaving no later than four in the afternoon!

Notwithstanding all these irritations and with ground engineer, pilot and flying field licences at last in place, Berkshire Aviation finally got underway with passenger flying at Thatcham, near Newbury on 27th May 1919. In its first season the small team, which included a mechanic named Blanshard, a rigger called Ritchie and Fred's wife Joan as seamstress in charge of fabric repairs, visited thirty-four towns. It usually stayed for one or two days, but remained much longer if the appearance coincided with a profitable 'wake' or holiday period.

Alan Cobham [left] with John [centre] and Fred Holmes [right] were the founders of Berkshire Aviation Company in April 1919. They are shown here with their first Avro 504K D9298 (G-EACL).

Alan Cobham (front cockpit) and Fred Holmes pose with passengers at Reading in June 1919 with D9298, prior to it being registered G-EACL.

Avro 504K G-EAIB was obtained by The Berkshire Aviation Company to replace G-EACL.

A major problem that soon arose was how to maintain the Avro's French-designed, British-built, 110hp Le Rhone engine without compromising the aircraft's daily earning ability. At Aylesbury, for example, this difficulty was overcome by working through the night in the open air, using the overhanging branches of an oak tree as a derrick in order to hoist out the engine. Improvising by the light of paraffin flares and electric torches, the ground support team then managed to complete the necessary overhaul before daybreak.

However, engine reliability continued to be a major problem and Cobham later recalled several instances when the possibility of paying farmers for crops damaged during a forced landing threatened financial ruin. At Northampton, on 21st July 1919, only a few days after a similar event at Leighton Buzzard, serious trouble loomed when a misfiring engine forced G-EACL to land in a hayfield off Houghton Road at Rush Mills and overturn. Cobham's initial dismay at the prospect of cancelled bookings was soon dispelled by the attitude of his two passengers, Miss Davies and Miss O'Callaghan, who, sympathising with his bad luck, expressed their eagerness to undertake another flight (and which in due course they did). Cobham's pessimism was further eased when he realised that, with the recent takings of between £500-£600 and a reasonable settlement from the insurance company, the purchase of a second machine was within the company's grasp.

Cobham caught the first available train to London and journeyed on to Croydon where he stayed overnight. First thing the following morning he presented himself at the Aircraft Disposal Board and soon selected Avro 504K D9303 (later G-EAIB) as a replacement aircraft. This machine, he was told, had already been earmarked for another customer, but the sight of ready cash not only clinched the matter, it attracted a £50 discount. Cobham immediately flew his new acquisition to Hamble where he persuaded the Avro team to convert it overnight into a three-seater. Returning to Northampton, only a little more than forty-eight hours after his departure, he found that Fred Holmes had an official inspector on hand to certify the aircraft. Berkshire Aviation was back in business!

More good fortune attended the Cobham-Holmes venture when the insurance assessors, unwilling to wade across a muddy river, nor to undertake a ten-mile detour to verify the claim, agreed to accept Fred Holmes' word that G-EACL was a complete write-off. Compensation of £500 was agreed, less £50 for salvageable spares, but when the machine was eventually turned over, a thorough inspection revealed little more than a collapsed front skid, a broken propeller and some minor superficial damage. After local repairs, Cobham flew it back to Yew Tree Farm where it was discreetly stored as a ready source of spares.

Berkshire Aviation's small team now began to travel further afield, but it was arranged that, whenever possible, return flights would be made to the home base at East Hanney so that bundles of laundry could be dropped on the lawn of the Holmes' family home, *"The Mulberries"* in Main Street for mother's urgent attention.

The quality of lodgings experienced during Berkshire Aviation's wanderings over the early post-war years varied enormously. Jack Holmes was fond of recounting the story of one small establishment in Somerset where, late each evening, in response to a knock on the door, he would open it to find a cup of hot chocolate outside – but mysteriously, no-one there to thank. On the last night however, he discovered his benefactor to be an elderly lady who explained that as she was sure he was going to kill himself, she didn't want to recall afterwards what he looked like!

The Holmes brothers and Cobham soon began to stay at establishments frequented by touring theatrical companies where they learned to appreciate more fully the value of proper business promotion and showbiz ballyhoo. Perhaps the best example of this came with their realisation that 'wing-walking' would give a new dimension to the stunt programme. So it was that at Leicester, on August Bank Holiday Monday, 1919, Jack Holmes and another member of the ground support team, R Graham-Wolland, introduced this 'daring sensational exploit' to an awe-struck British public.

The Berkshire Aviation Company also began paying more attention to advance publicity with offers of free flights in local newspapers. At Chesterfield, the whole company of a local revue, featuring the immensely popular comedian Will Fyffe, was persuaded to visit the flying field. The consequent publicity provided a great boost to the cash flow. Fyffe even encouraged the joy-riding team to travel north and to visit Edinburgh at Christmas. With wry humour, he confided that it was the only time the Scots could be parted from their money!

A 'staged' photo showing OP Jones wielding a hammer, flanked by mechanic Williams, Fred Holmes and Jones' wife at Barrow-in-Furness in February 1920.

But Cobham's meeting with the famed entertainer had a much more personal outcome. En-route to Scotland, the team arrived at Middlesbrough where the highly entertaining revue 'Joy Bells' was appearing at the city's main theatre. As at Chesterfield, the entire cast was invited to the airfield for free flights and it was during the course of its visit that Cobham became infatuated with the show's leading lady, Gladys Lloyd.

Keen to introduce her to the joys of flying, he was sadly disappointed to learn that she had already flown from Southport beach earlier in the year with Avro's exotically named Joseph Carey Crabtree Taylor. When Miss Lloyd's show moved to its next venue, Cobham, his 'love at (almost) first flight' undiminished, insisted on Fred Holmes driving him to Newcastle in order to be with her. His determination clearly paid off for, in less than a week, she accepted his offer of marriage although it was not until 30th June 1922 that the knot was finally tied, at St Giles Church, Bloomsbury.

In the remaining months of 1919, with public enthusiasm for the new flying game seemingly indicating a bright future, Cobham alone carried 6,000 people safely into the air. The fact that all this was achieved without serious mishap appears to have been largely a question of good luck if the experience of one man who, as a small boy attending Cobham's visit to Walton-le-dale, near Preston, is anything to go by. In a letter written to Cobham some forty years after the event, Mr TEAK Jackson, then the Principal of St Helens Technical College, recounted how, in an attempt to move the crowd which had overflowed onto the landing area, Cobham flew very low over their heads. He pointed out that his brother, over six feet in height, just managed to fling himself sideways, and that he, only three feet tall, merely had his cap removed by the undercarriage. It did however, acquire a long greasy mark and became a memento he treasured for many years!

In 1920, whilst also occasionally trading under the name Cobham and Holmes Aviation Co, and encouraged by promises of cash support for a more extended programme, Berkshire Aviation sold G-EAIB to Anderson and Pool Aviation, a joy-riding firm based at Priory Heath, Ipswich and purchased two more 504Ks G-EASF (D5858) and G-EAKX (H2600). Extra staff were recruited that included as pilot and 'wing-walker' the son of Peru's Minister for the Interior, Harold Gomez-Cornejo. They also took on Oscar Philip (OP) Jones who, two years later, left to become one of the country's best known civil Captains, firstly with Instone Airways and later with Imperial Airways and the British Overseas Airways Corporation.

Gomez-Cornejo's stay was, however, short and he departed in mid-1920 to join Captain JM Drysdale's Oxfordshire Aviation Company, then performing at Newport, Monmouthshire.

Jones' tenure with Berkshire Aviation was not without its memorable moments, for even his introductory flight with Alan Cobham involved a forced landing with repairs having to be carried out on a leaking fuel pipe with 'insulation' tape. A newsworthy event later occurred when Jones decided it would be appropriate to drop a large sheaf of flowers from the air during the unveiling ceremony of a Cenotaph at Stoke-on-Trent by Field Marshal Sir William Robertson. Praiseworthy though his intentions were, Jones' scheme backfired when the package fell onto the back of a policeman's horse, causing, as reported in the local press, 'a bit of a rumpus'. A prosecution for low flying soon followed. Fortunately further embarrassment was avoided when Police Inspector Adlem was persuaded to serve the summons in flight. This he did, with it duly endorsed "served personally in mid-air 24th November 1920". The 'adverse' publicity, it would appear, did Berkshire Aviation no harm at all!.

Unfortunately the company's early success heralded a false dawn and 1920 proved financially disastrous. Berkshire Aviation, which also traded in the early part of the year with G-EASF as Cobham and Holmes Aviation, found its cash resources impossibly stretched by trying to provide simultaneous shows at different locations. Problems brought about by continuous bad weather that often prevented flying, were compounded when the financial backing promised by local businesses at the start of the season also failed to materialise.

Facing bankruptcy, the company managed to survive only because certain sympathetic creditors refused to press claims. For

Cobham however, the writing appeared to be clearly on the wall and absorbing his share of the assets and debts, he resigned in May 1920 to reconsider his future, which, at that stage did not seem likely to embrace aviation.

Following Cobham's departure, an impressive list of pilots flew with the company which survived in various forms over the years 1921-1931. These included JCC Taylor, AL Robinson, AN Kingwill, JD Parkinson, CS Kent, GR Beck, L Leleu, RM Stirling, LJ Rimmer, H Lawson and B Ferrand. Another pilot was FGM Sparks, who, it is said by East Hanney resident, Nora Rickards, flew with a horn attached to his aircraft so as to announce his arrival!

During this ten year period, the company's fortunes picked up sufficiently for it to stay in business undertaking joy-riding, overhaul and conversion work. Though based initially at East Hannay, in November 1925, the company, now called Berkshire Aviation Tours, moved to the old First World War airfield at Witney, near Oxford and in the following year it became a limited company. During this period, more Avro 504Ks, including G-EBCK, 'FV, 'IN, 'KB, 'KR and 'KX provided a joy-riding service, though several moved quickly on to new owners. Berkshire's stay at Witney was relatively short and it took up residence at Monksmoor, Shrewsbury in December 1927 to conduct winter overhaul work before the following pleasure flight season.

But more reorganisation and relocation was soon to take place. In April 1929, Fred Holmes, with Manchester businessman, John Leeming, set up a holding company, Northern Air Transport Ltd to take over Berkshire Aviation Tours Ltd. The new organisation was then based at Wythenshaw before it transferred to Manchester's Barton aerodrome in June 1930.

With its fleet of sixteen red and silver painted 504Ks, the new company, with Fred Holmes as its administrative head and AN Kingwill as its chief pilot, visited many parts of the British Isles. Billed as 'The Great Air Pageant', its key performers at this time were the extraordinary stunt pilot Jock Mackay and multi-talented 'wing-walker' pilot, Martin Hearn. Both, as will be seen, were to contribute largely to the later success of Sir Alan Cobham's National Aviation Day Display tours.

Among the many aerial novelty acts featured in 'The Great Air Pageant' was 'The Flying Marksman' in which Colonel MU Stard

Berkshire used Yew Tree Farm at East Hanney for repairs and maintenance.

Lionel Leleu, relaxing between flights in Berkshire's Avro 504K G-EBKR, was one of several ex-Service pilots in the 1920s who went on to Imperial Airways. He was killed in March 1933 in the crash of Argosy G-AACI.

('Stiffy') – late of the Pyrenese Camel Corps and allegedly just returned from Neuralgia, demonstrated his uncanny skill at shooting and bursting balloons. Another 'wing-walker' was Tom Harriman who, as the 'Meandering Mechanic' was usually taken up by Jimmy Orrell, later destined to become chief test pilot at AV Roe. Another hugely popular act at many of the air displays was the arrival of a bridal party, supposedly departing on their honeymoon in a rickety old car, only to be bombed with flour bags from a low flying 504K by a losing suitor vowing revenge!

One of Northern Air Transport's fleet of silver and red Avro 504Ks, based at Barton, Manchester.

Berkshire's Avro 504K G-EBKB was withdrawn from use after its ducking off Scarborough on 13th September 1931.

One of Berkshire's regular stunts was 'bombarding the bridal couple with flour'; here seen at Enfield Air Carnival on 19th April 1931.

In 1932, Fred Holmes dissolved his association with Northern Air Transport, and formed Air Travel Ltd at Penshurst in Kent. Here, in addition to carrying out overhaul and conversion work, his new company provided a varied fleet of Avro and de Havilland machines for joy-riding. Air Travel later took up residence at Gatwick in the years immediately prior to the Second World War.

At this point, in order to enlarge the story, it is necessary to re-visit the periods just prior to and immediately following the First World War. One company that played a significant role in the development of pleasure flying was The Eastbourne Aviation Company Ltd. Originally formed on 18th February 1913 by the amalgamation of Major Frederick B Fowler's Eastbourne Aviation and the Frank Hucks Waterplane Co, it was based at The Crumbles, just east of the town. It provided a combination of flying instruction, joy-riding and aircraft manufacture and gave steady employment to some forty management and staff prior to the outbreak of war. This number increased significantly as production and repair work later expanded to meet Admiralty and War Office orders.

Although Eastbourne Aviation's main construction business ceased in 1919, six Avro 504L seaplanes were completed that uniquely embodied three separate cockpits. These were G-EAFB, 'JH, 'LO, 'NS, 'SD and 'SE which joined two Avro 504Ks, G-EAJG (H1956) and G-EALD (H1925), and two Short 184 four-seater seaplanes G-EALC (N2998) and G-EAJT (N2986), to form the company's post-war joy-riding fleet. Operations from the beach near the town's Royal Hotel were largely successful although set-backs occurred when G-EAJH sank off Hove on 19th August 1920 and two other machines, G-EANS and G-EALO were lost in crashes soon afterwards.

Throughout the peak period of 1919-1921, Major Fowler determinedly attempted to instil local interest in flying by organising air races and flying displays. He also opened a 'School of Flying for Ladies' which unfortunately did not fulfil its early promise. Inevitably, financial difficulties saw the run-down of the company and all work effectively came to a halt with the appointment on 23rd December 1922 of a local solicitor as Receiver.

Farther east along the coast, Major Brian Ferrand's smaller business also ticked over nicely with his Avro 504K/L G-EADK carrying 350 joy-riders from Folkestone's sea-shore during the summer of 1920.

The Navarro Aviation Company Limited and the Bournemouth Aviation Company Limited were two other companies that also entertained high hopes of cashing in on joy-riding's early post-war popularity. Joseph Navarro's 504Ks G-EADY, 'EA, 'EB, and 'JP enjoyed lucrative seasons at Whitstable and Southend and Bournemouth Aviation's assorted fleet which relied principally on its five 504Ks G-EADR, 'HK, 'RZ, 'SA and 'SB was in brisk demand as townsfolk and holidaymakers alike clamoured to fly over the South Coast. But neither company continued its joy-riding activities beyond the end of the 1920 summer season.

Flying from the seashore did however provide a popular 'no-frills' option that incurred minimum overhead costs for small concerns that operated around the British coasts. Although, remaining records are fragmentary, *"The Aeroplane"* magazine did make special mention of Mr Bernard Martin, who, equipped with a DH.6 G-EAQC, was awarded the flying rights for the 1920 season at Cleethorpes. Unfortunately, this aircraft was lost in a crash in November 1921. But, following this mishap, Martin expanded his operation to become the Martin Aviation Company, thereupon flying, in 1921, not only from familiar locations in Lincolnshire, but also from fields and beaches on the Isle of Wight with a replacement DH.6, G-EAWT. This machine, along with another two DH.6s, G-EAWU and G-EAWV, were purchased from the Brompton Motor Company and flew alongside Martin's Avro 504K, G-EAOE. This Avro too, however, was destined to have a short-lived passenger-carrying career, being written off in a crash at Cleethorpes on 12th June 1922.

Although the Avro 504 then available in such large numbers became the almost inevitable choice of the joy-riding promoters up

One of Eastbourne Aviation Company's Avro 504L seaplanes, G-EAJH is seen here at Brighton.

Major Brian Ferrand's Avro 504L G-EADK at Folkestone in 1920.

The young Sylvia Boyden (left) entertained crowds when civil flying was resumed in mid-1919 and became a leading parachute 'artiste' in the 1920s.

Joseph Navarro's Avro 504K still sporting its serial as an interim registration prior to becoming G-EAEB.

and down the country, a machine of an entirely different character appeared in this role over north London for a short period in 1919.

Frederick Handley Page, determined to get into the post-war airline business, had re-purchased a batch of sixteen O/400 heavy bombers back from the Ministry of Munitions. It was his intention that this type of aircraft, originally produced by his company for the Royal Air Force, should equip a new subsidiary, Handley Page Transport Ltd, based at Cricklewood. By the end of March, however, only four had been converted to carry seven passengers in the enclosed cabin with a further two or three squeezed into the air gunner's open cockpit in the nose. During the Easter weekend of 17-22 April, three of these aircraft D8350, F5414 and F5417 piloted respectively by Lt. Col. Sholto Douglas, Major Leslie Foot and Captain Geoffrey Hill, took some 800 joy-riders for half-hour flights.

In addition to joy-riding, demonstrations of the Calthrop 'Guardian Angel' parachute were also given by 'Professor' Newall and Miss Sylvia Boyden, descending from a height of 1200 feet – an almost unheard of novelty event at the time. William Newall had, but a few short years earlier on 9th May 1914, jumped from a Grahame-White Charabanc Mark X flown by RH Carr over Hendon, and gained the distinction of becoming the first person to descend by parachute from an aeroplane. Despite this and later demonstrations, the military authorities decreed that such life-saving devices would undermine the aggressive spirit of wartime aircrew, and thereby condemned many British fliers to a needless death.

Resuming his jumping activities after the war, Newall gave many displays before suffering a fatal accident in Denmark in October 1922. On this occasion, Newall's parachute had caught in the aeroplane's tailplane. Whilst the pilot descended and flew low enough to allow Newall to struggle free from his harness and fall into the water to be picked up by a boat, the effects of the cold and exhaustion proved too much.

Miss Boyden too, had already secured a claim to fame when, aged just seventeen, she became the first woman to leave an aeroplane by parachute. She went on to become Britain's premier 'parachutiste' and, after having survived the crash of HP O/400 G-EAAF near Carlisle on 12th May 1919 in which she lost several front teeth, later accompanied Major Orde Lees (with a chaperone) on a tour of America to demonstrate the 'Guardian Angel' parachute.

Another large aircraft that featured briefly on the post-war joy-riding scene was the ungainly Blackburn Kangaroo, twenty of which had been originally constructed for wartime anti-submarine patrol duties. In May 1919, most of these were re-purchased by Blackburns from the Aircraft Disposal Board but three, B9981, B9982 and B9985, were acquired by the Grahame-White Air

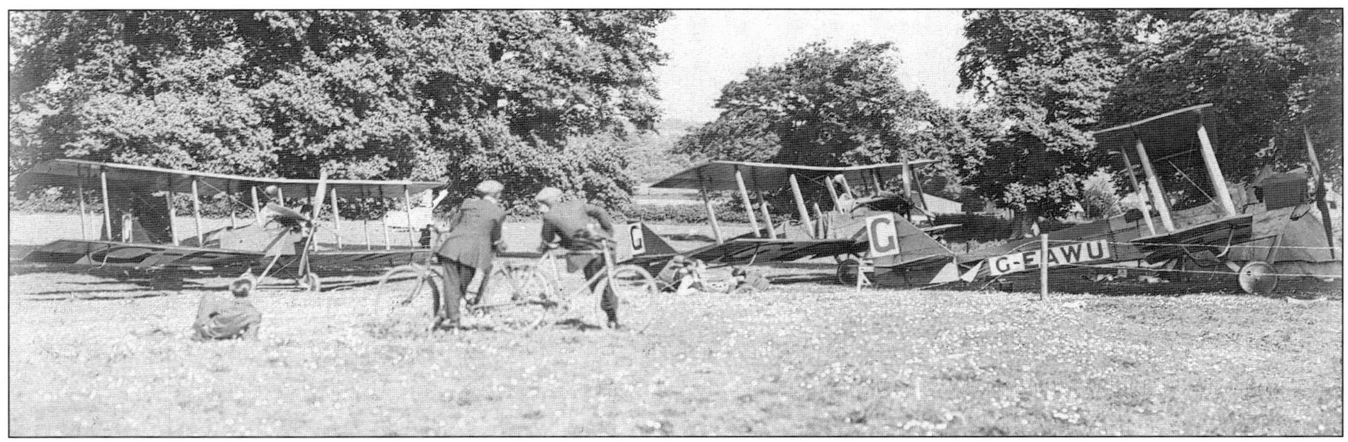

Three DH.6s of Martin Aviation Company 'somewhere in the Isle of Wight' in 1922.

Blackburn Kangaroo B9985 [later G-EADG] was one of three used by Grahame-White for joy-riding at Hendon. Note the Hucks starter.

Service to supplement its joy-riding fleet of twelve Avro 504Ks. After their arrival at Hendon, the three machines were stripped of military fittings and their fuselages modified to provide two separate cockpits that seated seven passengers. An eighth person was carried in the nose gunner compartment.

It was not until 8th June 1919, that civil registrations were allocated by which time B9982 (G-EADF) had already met with

Two well-dressed young aviators awaiting their first flight in Grahame-White's Avro 504K B8758 (G-EABH).

misfortune. This occurred when a wing dug into the ground following an engine failure during take-off at Hendon on 31st May 1919. Fortunately none of the joy-riding passengers was injured, but the mishap caused much concern at a formal reception being given there at the time in honour of the two national heroes, Harry Hawker and Commander MacKenzie Grieve, recently plucked from mountainous seas after their failed attempt to become the first airmen across the Atlantic. Only a few weeks later, on 29th June 1919, B9981(G-EADE) was also damaged beyond repair, leaving G-EADG to carry on alone undertaking pleasure flights and display work until its own withdrawal from service in mid-1921.

At nearby Stag Lane aerodrome, the London and Provincial Aviation Company Ltd also conducted joy-riding flights during this period with its L and P School Biplanes. In July 1919 however, the company, founded in October 1913 forsook aviation to enter the furniture and chocolate making businesses!

On 23rd April 1919, Robert Blackburn formed a subsidiary company, The North Sea Aerial Navigation Company to operate commercial services with three Kangaroos B9972, '73 and '78. Joy-riding soon complemented freight carrying and two Avro 504Ks, G-EAGV and 'GW were purchased to join Kangaroos G-EAIT, 'IU and 'KQ that operated initially from Roundhay Park at Leeds and later from Brough. The First Air Traffic Exhibition, known as ELTA, held in Amsterdam in August 1919, proved a highly successful venue for the three Blackburn machines. During the exhibition period, they carried between them some 1,400 fare-paying joy-riders and gave many VIP complimentary flights without incident.

Following the failure of the company's commercial operations in 1920, joy-riding became the mainstay activity for North Sea until an RAF Reserve School came under its management at Brough in 1925. G-EAIT suffered a fatal crash at the aerodrome on 5th May 1925, but three more Kangaroos, G-EBOM, 'PK and 'MD christened *"Pip"*, *"Squeak"* and *"Wilfred"* after the cartoon characters then featured in the *"Daily Mirror"*, continued on in the training role until withdrawn from use in 1928.

AV Roe was another company which, like Handley Page and Blackburn, having shrewdly re-purchased a large number of its own ex-wartime products, found itself well placed to enter the organised pleasure flight business once the wartime restrictions were lifted. At its Hamble facility, three 504K machines piloted by GLP Henderson, HA Hamersley and F Warren Merriam carried 359 passengers paying £1 each on the first day of the 1919 Easter holiday.

Encouraged by this, within one month, an off-shoot company, the Avro Transport Company, was set up at Birkdale Palace aerodrome near Southport and many leading pilots of the day were recruited and based throughout the country to fly Avro 504K and L (seaplane) variants alongside the five-seat Avro 536s. An Avro 534 Baby, (K-131) G-EACQ was also employed to publicise the joy-riding fleet by giving aerobatic displays along the South Coast. This machine was to have a chequered history for, after a crash on 13th January 1920 into the garden of the Rev. Everard Verdon Roe's Hamble vicarage, which nearly took the life of its newly qualified pilot, the future Avro chief designer, Roy Chadwick, it was rebuilt and sold to HJ 'Bert' Hinkler. It then went on in May 1920, to make a record-breaking 650 mile non-stop flight from Croydon to Turin in 9½ hours, for which Hinkler received the Britannia Trophy.

Avro's new organisation was broadly divided to cover 'Southern' areas (Hounslow Heath, Manston, Southsea, Weston-super-Mare, Brighton, Swansea, Paignton and Porthcawl) under Captain Duncan Davis and 'Northern' counterparts (Blackpool, Southport, Fleetwood, Rhyl, Liverpool and Manchester), with GLP Henderson in charge. Much attention in the north was paid to Wakes Weeks, where aerial attractions proved especially popular with the large crowds of industrial holiday-makers.

Clearly the success of the pleasure flight operators was highly dependent on favourable weather. It therefore came as a severe blow when, in mid-August 1919, ferocious storm conditions caused much damage to the Avros optimistically pegged down with sandbags on the beaches.

It was also during August that nine Avro Transport 504s joined company with the Blackburn Kangaroos and attended the afore-mentioned ELTA exhibition in Amsterdam. As with the larger Blackburn machines, the sustained demand for joy-rides proved highly profitable and it was not until 3rd January 1920 that the last two Avros returned to England via Lympne.

Avro Transport also established joy-riding centres at Douglas, Isle of Man and at Cockshot Point, Windermere, where Howard Pixton, winner of the 1914 Schneider Trophy competition, took up residence as chief pilot. Pixton later formed the Lakes Motor and Seaplane Company to take over the Windermere facility after the Avro Transport Company ceased operations in 1920. The Douglas franchise was taken over by the International Aviation Company based in Liverpool, but this soon failed.

Noting the success of Avro Transport's pleasure flying fleet at Southport, ex-military pilot Norman Giroux, who had flown for the International Aviation Company, set about forming the Giro Aviation Company in 1920. Despite the failure of many other like-minded concerns, his successful operation from Hesketh Park Aerodrome alongside Southport beach extended, with the exception of the wartime years, until well after the Second World War.

Avro Transport Company's D6230 [later G-EADX] seen here in somewhat cheerless conditions on Southport beach.

Avro's Lake Windermere operation employed Avro 504K/L seaplanes H2581 and H2582 [later G-EADJ & G-EADK] but was short-lived, ending in October 1919.

Initially using five Airco DH.6s G-EARC, 'VG, 'RK, 'RM, 'RJ, supplemented later by G-EBEB and 'WG and, from 1931, three Avro 548s, G-EAFH, G-ABMB and 'SV, he later employed an Avro Avian, G-EBZM and two DH.83 Fox Moths G-ACCB and 'EJ, sporting distinctive white fuselage and silver wing paint schemes. Giroux gained local fame following his dramatic rescue after a forced landing on the River Ribble in August 1921. Having managed to nurse his powerless DH.6 G-EAVG safely on to an exposed sandbank some two miles off-shore, Giroux and his two passengers, Captain Gibbons and Mr Davies, had saturated their discarded clothing with petrol and set fire to it hoping to attract attention, but without success. Giroux, being a strong swimmer, then struck out for the shore, but found the strong currents carrying him toward the mouth of the estuary. With hopes fast fading, he was fortunately spotted by the crew of a small dredger, Hopper No.87, whereupon he directed it toward his stranded passengers. With the tide rising fast over the sandbank, the rescue was achieved with little time to spare.

Giro Aviation's ability to keep going for so many years was exceptional. Many other early operators such as Blackpool Flying Services, Golden Eagle Aviation at St Annes-on-Sea, Kingsford-Smith Maddocks Aeros, Oxfordshire Aviation Services at Caversham, Welsh Aviation, Frank Neale's Essex Aviation, Warwick Aviation, West of Scotland Aviation and the Kingwill and Jones Flying Co with its headquarters at Alloa's Old Aerodrome were examples of pleasure-flight operations within the UK that ceased trading within a year or so of starting up.

Some companies put up a determined fight for survival. HW Barrett's Midland Aviation Service, based at Alexandra Park, the birthplace of the Avro 504K, operated G-EADP, 'GB and 'LE along with DH.6 G-EARR for a season before being taken over in 1922 by the Manchester Aviation Co Ltd. This concern, together with Manchester Airways, then provided joy-riding attractions throughout Lancashire before finally offering three Avros, G-EADP, 'ZW and 'ZX to found the Northern Aviation Co Ltd in 1924.

Leatherhead Aviation Services Ltd was created in 1920 by William George Chapman who, having progressed through the cycle and motor business, had taught himself to fly on a Deperdussin before the First World War. Having purchased a DH.6, G-EANU (C5230) and four 504Ks, G-EADU (D9329), G-EAHL (E4118), G-EBAV (F9783), G-EBCQ (H6547), Chapman engaged RC Knowles as his full-time pilot and a number of others as part-time flying staff. These included AF Muir, FH Hayns, LG Robinson, A Graham, SF Woods, FJ Ortweiler and FWJ Grant. Leatherhead's pleasure flight operations were mainly centred at Byhurst Farm, Maldon Rushett, near Chessington and Croydon aerodrome and frequently featured a French lady parachutist, Mme J Finet. On 20th August 1922 Chapman was badly injured when his Avro 504K G-EBAV, piloted by Arnold Graham, crashed in the grounds of St Bernards Convent at Slough. This marked the end of Leatherhead Aviation Services although Chapman did recover to continue a long career in the aviation industry.

Croydon also proved a happy hunting ground for Aircraft Transport and Travel Ltd which maintained a fleet of Avro 504Ks, G-EAIO – 'IS for joy-riding. This company's operations were by no means confined to southern England for its team, with Captain Sturt in charge also undertook pleasure flights throughout the summer of 1919 from the Stray in Harrogate.

At this time, another manufacturer, Vickers Ltd also catered for an enthusiastic public demand with Avro 504Ks, G-EACV, 'DS, 'EY and 'EZ based at Joyce Green in Kent and also at King's Heath, Birmingham. Meanwhile, on the east coast, some mystery attaches to the post-war appearance of Avro 504K, G-EASG at Bridlington's Sands Lane estate. Originally constructed as G-EAFX (H.6543) this machine was written off following gale damage in August 1919. It was rebuilt as 'SG and registered with Golden Eagle Aviation and later Lytham Aviation Co, but unfortunately records do not confirm that this aircraft operated in Yorkshire.

No better illustration of the 'fly today – gone tomorrow' outfits that became commercial casualties can be provided than that of Border Aviation Ltd. Formed in 1920 by Percy Ingham and two brothers, RF (Bob) and Graham Little, Border Aviation Co Ltd engaged Captain John Oliver to fly its Avro 504Ks G-EANQ (J755) and G-EAIA (F8717) from Botcherby near Carlisle and later from a field near Scarborough's Oliver's Mount. But the company's non-executive directors, concerned at what they regarded as an excessive number of mishaps, called for its affairs to be wound up at the end of the season.

Notwithstanding this, Border's assets were immediately transferred to the Little brothers who, along with Percy Ingham, re-constituted the outfit as the Ingham and Little Aviation Company. Major Ferrand who, during 1920, had flown his float equipped

Avro 504K G-EADP served with several joy-riding concerns before joining Northern Aviation in 1924.

Displaying an advertisement for a brand of soap, Avro 504K G-EASG is seen here at Sands Lane at Bridlington, although it was registered at the time to Lytham Aviation Company.

Border Aviation's Avro 504K G-EAIA providing pleasure flights at Scarborough and still showing its service roundels on the upper wings.

504Ls from the South Coast, also joined the new organisation flying G-EANQ at Heysham and Morecambe during the summer of 1921. However, the crash of this aircraft on the banks of the Solway Firth on 1st December while operating from Cockermouth meant that the company simply collapsed in the same way as its predecessor. Ferrand duly moved on to the Berkshire Aviation Co.

The fate of Border Aviation was also mirrored by that of Stallard Airways. Set up as a joy-riding operation at Penshurst in Kent with Arthur Boorman, Ronald Levy and Tom Baden-Powell as pilots, the company's three Avro 504s G-EAFQ, (E4180) 'SK and 'JG (H1956) and a DH.6, G-EANJ, were a common sight in the Kent and Surrey skies in 1921. However, by August 1922, a series of serious misadventures brought about the demise of the company, and indeed the closure of the airfield, although it was re-opened for private flying in 1927. In 1930, Ronald Waters and his company, Home Counties Aircraft Services, arrived at Penshurst. Operating two DH.60 Moths, G-EBWX and G-AAYF, it was also imbued with the noble intention of bringing flying well within the reach of every man and woman. Although initially encouraged by a season which embraced well over one thousand pleasure flights, Waters found it necessary later in the year, to relocate his operation to Gatwick, where the public interest was, unfortunately, not sustained.

At Northolt aerodrome in north London, the Central Aircraft Company employed Captain Anderson, HH Maddox, EB Wilson and Herbert Sykes as chief pilot, to provide pleasure flights over London, the Thames Valley, the Solent, Kent and Southend throughout 1919 and 1920. The company however ceased trading in 1921 and its fleet of eight Avro 504Ks was soon disbanded. Following this, Sykes, advertising himself as the 'Psyche's Circus', joined the band of solo operators in Avro 504K G-EACA chasing by now a less enthusiastic public. He was to find, as did so many other hopeful individuals that, in the wake of the 'boom and bust' period that followed the First World War, individual joy-riding operations generally failed to prosper.

Success was more often reserved for fewer but better organised teams. One such organisation was Surrey Flying Services created at Croydon in 1919 by Captain AF Muir and FWJ Grant with two Avro 504Ks, G-EAWI and 'WJ. This concern became Surrey Flying Services Ltd in 1922 and was joined by the Little brothers in 1923. Surrey progressed with a greatly expanded mixed fleet and undertook, in addition to joy-riding, both flying training and air-taxi work. During the 1920s and early 1930s the emphasis on these different aspects of the business fluctuated, as of course did the number of flying staff engaged by the company. During this period many well known pilots, such as Jimmy Youell, GLP Henderson, JC Chamberlain and ED Crundall transported thousands of thrill-seekers from southern county airstrips for five-minute "five-bob" (25p) "flips" in the company's extensive fleet that included many blue and silver Avro 504Ks and later, when these were retired, Clerget-powered Avro 536s.

At that time it was not unknown for private individuals to hire a Surrey Flying Services machine for a short spell only for it to be later discovered that, after having had a suitably impressive

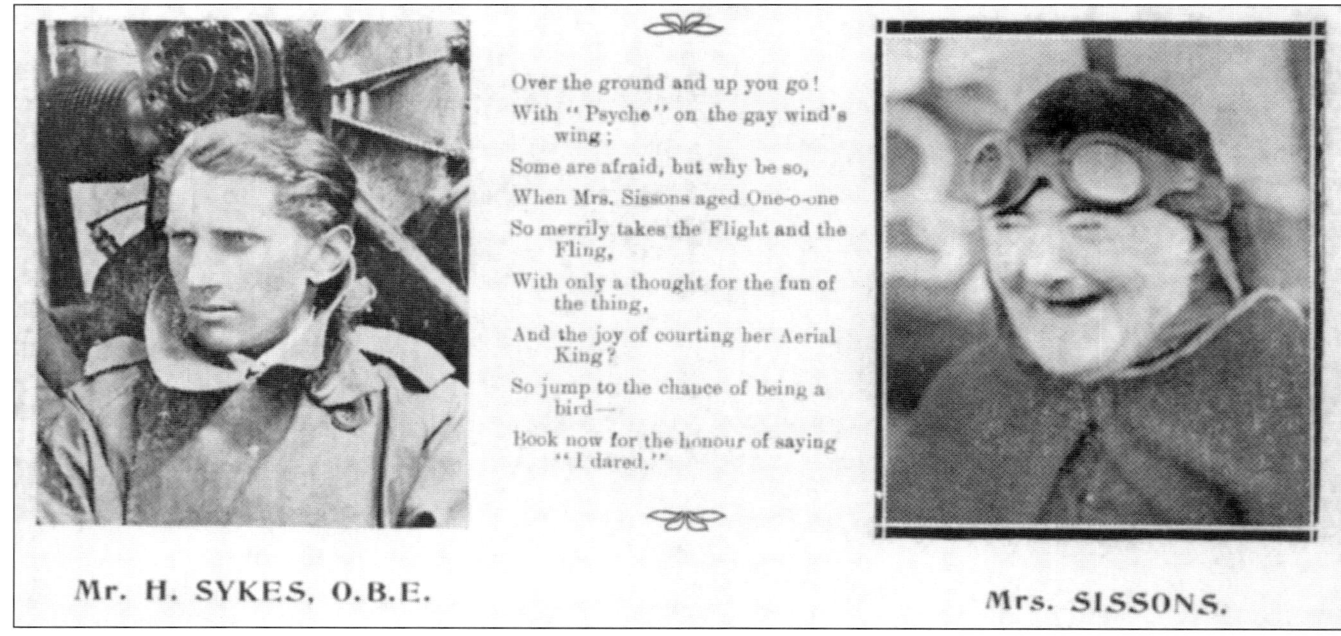

A typical souvenir postcard shows Herbert Sykes of Psyche's Circus with centenarian Mrs Sissons - Sykes was one of the many operators trying to cash in on the joy-flight boom.

'company' sticker affixed to the fuselage, it had actually been used for unregistered pleasure-flights at some seaside resort! It was also not uncommon, for the presence of a 'Surrey' machine to be challenged by a smaller competitor. A good example of this took place in 1929 when, at Southend, Albert Adam's Aeroplane Services Company pitted an Avro 504K, G-AAEZ against Surrey's G-EBYW in the fight for fare-paying passengers.

In Yorkshire, it was in a farmer's field off Endike Lane, Hull, that Amy Johnson, later to become Britain's best known aviatrix, took her first flight in one of the Avro 536s converted by Surrey Flying Services to carry four passengers (almost certainly G-EBOY). Other districts attended by 'Surrey' included the Isle of Man and the Channel Islands where at Jersey, flights took place from the beach.

Burry's Farm at Somerford Bridge near Christchurch (then in Hampshire but now in Dorset) was also visited by Surrey Flying Services. This subsequently became the home of Captain Francis Colbourne Fisher's joy-riding 504K, G-EBVL, during the years 1931-1934. Fisher then went on with Havelock Clive-Smith to form, on 31st August 1934, Fisher Aviation Co Ltd, principally to develop the site into a recognised airport. Though faced with much local opposition regarding aircraft noise and the likelihood of devalued property prices, he persevered, and with Sir Alan Cobham as chairman and Fisher as managing director, what became formally known as Bournemouth Airport Ltd came into being in February 1935. However, this site is not the present Bournemouth International Airport, the present home of FR Aviation Limited (a major operating company within Cobham plc), and which was created in the Second World War as RAF Hurn.

At the end of the 1920s, public interest in joy-riding, which had bloomed and then faded just after the Great War, once again began to re-awaken.

Surrey Flying Services operated joy-riding at Croydon Aerodrome for many years after the Great War; here is their first Avro 504K G-EAWI in 1921.

In 1930, Freddie Kent, who had left Imperial Airways some three years previously to become Surrey Flying Services' chief engineer, was enticed back by his previous employers to accompany the company's Handley Page W.8 airliner, G-EBBI, 'Prince Henry', on a seven month publicity tour of the country. The road show, operated by Aviation Tours Ltd of Croydon, and with its own director Captain Earl B Fielden, piloting *"Prince Henry"*, commenced on 4th April with a ten-day stay at Port Meadow, Wolvercote situated some three miles from Oxford.

"Prince Henry" then undertook joy-riding visits to over thirty major towns where its arrival, when possible, was arranged to neatly align with another significant aerial event. At Hull for example, Aviation Tours' presence at Hedon Airport coincided with Amy Johnson's rapturous homecoming after her famous flight to Australia in 1930. While the costs of the tour were largely offset by income generated from pleasure flights, the management, equally mindful of not wasting any time when bad weather precluded flying, charged the public sixpence (2^1/$_2$p) to inspect the machine.

After this tour was completed, Kent joined Rollason Aviation, a concern started in 1930 by Captain William A Rollason and co-director Walter F Rickard which had a varied fleet of aircraft based at Croydon for flying school instruction and air charter work. In March 1932, the management team was further strengthened by AF Muir and the company was re-formed as Rollason, Muir and Rickard. Profits from joy-riding using a Desoutter Mk II, G-AAZI encouraged the owners to increase the number of machines. As will be seen, it was a Rollason-owned Desoutter Mk I, G-AANE and a three-seater Spartan G-ABWV that, in the capable hands of display pilots Tommy Nash and 'Jock' Bonar, accompanied Sir Alan Cobham's National Aviation Day tour during the 1932 summer season. The following year, both these aircraft, along with a ten-seater DH.84 Dragon G-ACCR, flown by Captain Rollason himself formed part of the impressive British Hospitals Air Pageant that, led by the charismatic Charles W A Scott, also toured Great Britain.

In May 1924 another company was formed which, as with Sir Alan Cobham's later enterprises, would become synonymous with the 'flying circus' and pleasure-flight business. The Cornwall Aviation Company of St Austell was founded by Leonard Hill and Captain Percival Phillips after the Berkshire Aviation Company used Phillips' garage business for its engine overhaul work. It appears that Phillips, invariably referred to locally as 'PP', needed little persuasion to return to a life of flying. Firstly, he acquired an Avro 504K, G-EBIZ for £250 from Berkshire Aviation. This aircraft, instantly recognisable in the company's bright red colour scheme went on to carry some 91,000 passengers over the next ten years. In so doing, it became Britain's most famous joy-riding machine. Phillips, having recruited Albert J

in 1930, Freddie Kent was in charge of engineering matters with Aviation Tours' HP W.8 G-EBBI.

Adams as his licensed ground engineer, wasted little time in acquainting the Cornish resorts with his superb acrobatic skills – not neglecting to take Don, his ever-willing canine companion with him whenever the opportunity arose. Equally at home on either the pilot's lap or the floor of the rear cockpit, Don, like Captain Phillip's young son David, survived many extreme manoeuvres totally unrestrained by any kind of safety device!

There was no shortage of passengers – women frequently outnumbering the men – willing to pay five shillings for a short circuit of the airfield, fifteen shillings for a flight including a loop or a pound for a spin. 'PP' enjoyed a highly successful first season despite constant parochial opposition to his flying on the Sabbath. This was a recurring complaint that, coupled with other objections that the low flying and 'unendurable' noise levels affected the elderly and infirm, was frequently encountered by joy-riding operators throughout the country.

It was, however, going to take more than minor issues such as this to dampen the spirit of 'PP' who had clearly found his niche. In 1925 he set off on a tour of the West Country that included an extended stay at Paignton. The following year, having recruited Captain EW Jordan and purchased another Avro 504K, G-EBNR, Phillips, in addition to visiting many towns in the Home Counties, performed at several favourite locations along the South Coast. These included Weymouth, Folkestone, St Leonards and Margate where he used a field at Nash Road near Chapel Hill. A car stationed at the clock tower on the seafront transported a constant supply of passengers to and from the flying field.

Desoutter Mk I G-AANE of Croydon-based Rollason Aviation was one of the aircraft they flew with the Cobham and Barker-McEwen King display tours.

Another 'happy handful' from the 91,000 passengers carried in Cornwall Aviation's famous red Avro 504K G-EBIZ.

An incident that received much publicity occurred in July 1927 when Captain Phillips was called before the local magistrate to answer an accusation of flying dangerously low over Margate beach. According to one police witness, Phillips had failed to maintain a safe height, though he admitted this had been viewed from a distance. Another piece of 'damning evidence' stated that a lady was caused to faint when Phillips' machine 'swooped only three yards above her head'. Fortunately a photograph showing enthusiastic bathers waving to Phillips as he flew by well offshore, coupled with his reputation as an experienced and highly decorated wartime pilot, quickly established that there was no case to answer. The 'King of the Joy-Riders' was then immediately cleared to continue operations.

By this time, 'AJ' Adams, in addition to the role which he shared with PB Swan and FE Draycott as 'aerial aerobat' (ie 'wing-walker' and stuntist), was also being called upon to find more suitable landing fields for Air Ministry approval. Fortunately the system, which had proved such a bottleneck when first required by Berkshire Aviation in 1919, had improved to such an extent that applications were now routinely approved following the mere submission of a site marked on a six-inch ordnance survey map with Phillip's signature alongside. Such was the regard in which Phillips and Fred Holmes were now held that they were authorised to represent many other small joy-riding firms for the licensing of temporary landing grounds.

Supporting his defence against a claim of dangerous flying, Capt Phillips is shown to be well offshore at Margate in July 1927.

Cornwall's team seen here with G-EBIZ at Paignton in September 1925 are [left to right] Mike Crocker, Gerry Draycott, Captain Percy Phillips and Mr & Mrs AJ Adams.

The Cornwall Aviation Co., Ltd., St. Austell

present THE

FLYING CIRCUS

KEEP THIS PROGRAMME

FLIGHTS FROM 5/-

DAILY—10 A.M. TILL DUSK

A BURLESQUE AIR DISPLAY
SOMETHING **NEW** AND **DIFFERENT.**

HANHAM—3rd to 14th JUNE

(at Tram Terminus)   Flying All Day—Noon till Dark

"THE FLYING FOOLS"
will **THRILL & AMUSE** you ALL the time.

DISPLAYS NIGHTLY at 7.30 Weather permitting

"BOMBING OF CHARLIE CHAPLIN"
Great Fun for the Kiddies — Bombs-Stick-Cycle-Hat—Oh! what a Mix up.

Stunt Flying ════════ Crazy Flying

Looks very thrilling—It isn't really, so we are giving two of **you** a chance to prove it —**FREE.**

Programme liable to Alteration in event of unfavourable weather conditions

EXACTLY! — We call him "Leonie." You will like this fellow.

SIMULTANEOUS WING WALKING

Just that—All very simple

"The Ungrateful Passenger."

Well—He Got What He Asked For.

SURPRISE FLIGHTS GIVEN AWAY EACH EVENING
Admission 6d. Children 3d.

THE FLYING CIRCUS
(Weather permitting) present their

SUPER AIR DISPLAYS (TWICE DAILY)

ON SATURDAYS & SUNDAYS AT **3-45** & **7-30**
Full of Thrills :-: Full of Fun.

THE FULL WEEKLY PROGRAMME and in addition.

FORMATION FLYING
Our Pilots don't like this a bit—still, why worry—you will.

AERIAL SHOOTING
This is clever—providing everything GOES OFF alright.

"AUNTIE TRIES FLYING"
She does her best—but—

"THE FLYING SQUAD'S UNLUCKY DAY"
Ends in Smoke—What Fun— For You—Not Them.
Laugh at the Police with Safety.

ALL Bristol Should See This Humourous and Unique Display
FAMOUS PILOTS :: MUSIC :: REFRESHMENTS

ADMISSION 1/- (Including Tax)   **CHILDREN 6d.**
CAR PARK

B. H. PRIEST, Printer, 22a & 24 Balmoral Road, Westcliff-on-Sea.

This was followed up with a proposal, endorsed by Sir Sefton Brancker, then the UK's Director of Civil Aviation, that other selected pilots with 250 hours of operation from this kind of small restricted landing field, should also be given delegated approval – on the understanding that formal paperwork would be submitted soon afterwards. However, following Brancker's tragic and untimely death on 5th October 1930 in the R101 airship disaster, this proposal was rejected by the Air Ministry.

With the growing success of Cornwall Aviation, the need arose for a third aircraft and additional pilots. In 1927, Avro 504K G-EBSE was acquired and Sammy Summerfield and Captain Don 'Jock' Cameron joined the company – thus allowing operations to extend into the northern counties and Scotland. The traditional rivalry that exists between Yorkshire and Lancashire, best exemplified in the 'Roses' cricket encounters, was never stronger than in the inter-war period. It was therefore perhaps not surprising that some Lancastrian feathers were considerably ruffled when, on 27th September, the Yorkshire Aero Club was invited to give a flying display at Morecambe's Carnival. Local indignation at the 'Tyke' involvement was only cooled when the invitation was withdrawn and non-partisan Cornwall Aviation was persuaded to field an alternative crazy flying team – albeit for what turned out to be a thoroughly wet occasion!

Cornwall's flying strength was further increased in 1928 when 504K G-AAAF was acquired and at the same time, G-EBIZ was converted to a four-seater. No doubt influenced by Phillip's success, 'AJ' Adams decided to break away and, in March 1929, with Avro 504K G-AAEZ, started up his own joy-riding and aerial advertising firm, Aeroplane Services Ltd in Birmingham. Three other 'Cornwall' employees left to accompany him including pilot C Bint, but what has been recorded as 'an unwillingness on Mr Bint's part to persevere with the effort' soon rendered the venture unviable. By the end of the year, the company had gone into receivership and Adams had returned to become Phillips' business manager.

With so many take-offs and landings taking place, occasional mishaps, fortunately mainly of a minor nature, occurred on Cornwall's operations. One particular example took place in 1931, when Captain Crundall, late of Surrey Flying Services and now flying with 'PP', had the misfortune to experience engine failure when flying over Weston-super-Mare. Left with no option but to land quickly on the shore, he did so, but his abrupt arrival caused the passengers to be catapulted out of the aircraft and into the mud that adjoins the seafront. Needless to say, the company's claim that its previously unblemished record at the town with fifty uneventful flights a day over the preceding two years, made little impression on local officials.

Above Left: Cornwall Aviation introduced novelty acts in a Burlesque Flying Display at Hanham, near Bristol in June 1931.

Left: A typical 'made-up' souvenir postcard which shows Albert Adams whilst he was at Cornwall Aviation.

Cornwall's 'crazy gang' included Albert Adams [extreme left], Captain West [tartan suit], Frank Cradock [white overalls], Bill Craig [as Charlie Chaplin] and Capt Crundall [seated as 'Aunty'].

Chapter Three: A New Awakening

*'High Spirits they had,
gravity they flouted'*

Cecil Day Lewis

By 1931, Cornwall Aviation with G-AAYI now also added to its 504K fleet, had carried 95,000 people and 'PP's' claim to have taken up 55,000 (more than any other pilot in the country) was undisputed. Nevertheless there was no denying that the public's enthusiasm for joy-riding with small operators was again diminishing. Clearly something needed to be done. It was Albert Adams who, remembering the flying circus performers he had seen when living earlier in America, suggested that Cornwall Aviation should introduce similar features into its touring repertoire.

Phillips' team then commenced its new Burlesque Flying Display with a two-week stay at Hanham near Bristol in June 1931. Bad weather seriously hampered the proceedings with floods, hailstorms and a minor earthquake preventing many joy-riders, including several bus-loads of children who had won local newspaper competitions, from getting off the ground. Nevertheless, 'Aunty' Crundall and general factotum Bill Craig, masquerading as Charlie Chaplin, managed to entertain the crowds with comedy routines. These included the dropping of flour bombs onto a 'criminal' hotly pursued by 'police'. But the risk involved in dropping any object from the air was well demonstrated when Len Clemo, one of the ground assistants at a venue in Taunton, suffered a direct hit by a flour bag. He spent several uncomfortable days in hospital with severe bruising.

The popularity of such pranks encouraged Phillips to consider a similar undertaking in 1932. However, sponsorship was by now proving difficult to obtain so, with two more 504Ks, G-ABHI and G-AAUJ now complementing his fleet he agreed to pay 30% of his takings to work as a 'freelance' alongside other display entrepreneurs, most notably, as will soon be seen, Sir Alan Cobham, then about to launch his National Aviation Day Display campaign across the country.

The period from 1931 to 1937 saw the 'high summer' of large-scale displays that, led by well known aviation personalities, allowed the general public to actually take part in the flying events. One such celebrity was Captain Charles Douglas Barnard, who, like Cobham, had made a name for himself in the post-war flying fraternity as one of Captain Geoffrey de Havilland's Aeroplane Hire Service pilots. Latterly he had achieved public recognition by accompanying the Duchess of Bedford as her private pilot on several long-distance flights across the world in her Fokker F.VIIA G-EBTS, *"Princess Xenia"*.

Having now purchased this aircraft for £900 and renamed it *"The Spider"*, Barnard, along with his wife Matilda as co-director, set up CD Barnard Air Tours Ltd. Sponsored by the *"Daily Mail"*, his team, frequently advertised as the 'Barnardstormers', spearheaded what was to become a revived public interest in aviation, visiting 118 towns between 1st April and 11th October 1931. A varied collection of aircraft which, led by Barnard himself in *"The Spider"*, included Spartan three-seaters G-ABJS and G-ABET flown by ED Ayre and FS Crossley, a Cierva C.19 Autogiro, G-AAYP piloted by Captain Reggie Brie, and, most unusually a Potez 36 Ladybird, F-ALJC (G-ABNB, registered to the Earl Halsbury) that was demonstrated by Bud Cummings. An Avro Avian, G-AAXH, flown by Len Stace, Jim Mollison's DH.60X Moth and Joe King flying a Desoutter owned by National Flying Services completed the fleet which Barnard fairly claimed to be the first of the major touring displays to travel entirely round the country.

Though conducted without major mishap, the tour came close to experiencing serious trouble on two occasions. The first occurred at Glasgow's Renfrew airport when Neville Stack's Spartan, left unattended with the engine throttle friction nut untightened, moved off into the path of the taxiing Avian, knocking

Captain CD Barnard with his famous Fokker F.VIIa G-EBTS "The Spider" undertook display tours in England and India in the early 1930s.

off its right-hand wings. Gathering pace, the Spartan then continued towards the spectators but, just before reaching the fence, it ran over uneven ground and pitched onto its nose.

While at Stirling in Scotland, only Mollison's skilled airmanship averted another potential disaster. Flying Stace's Avian, he experienced an engine cut at some 100 feet but managed to overcome a critical situation when, turning to land his gliding aircraft, he threaded his way between two Spartans already committed to take off in the opposite direction.

Following the 1931 tour, Barnard went on to feature *'The Spider"* at air gatherings around the country such as that organised by Thanet Aviation Ltd at Nethercourt Aerodrome, Ramsgate on 20th-21st August 1932. Thanet Aviation had itself entered the 'five bob flip' market with Ernie Bicknell, its chief pilot, operating an Avro 504K G-ABJF at Ramsgate and PH Meadway another, G-AAWC, at Gravesend.

In 1933-34, Barnard's organisation, now re-formed as India Air Pageants Ltd, and managed by Captain Arthur Dalton visited the Indian subcontinent, giving its first display on 23rd December 1933 at Juhu Aerodrome, Bombay. The tour, with *"The Spider"* flown by Barnard in company with nine other assorted machines also featured three of Britain's best known parachutists. The tour proved to be an unmissable attraction for countless thousands, who, in all probability, may never have seen an aircraft before; although Nevill Vintcent and John Newall, calling themselves the 'Air Tramps', made a limited tour of India in 1928 in a pair of DH.9Cs, G-EBUM and 'UN.

One of the pilots recruited to fly in the Indian display was ex-RAF Flight Lieutenant GC Sullivan. He had set out from Britain in Captain Dalton's Fox Moth, supposedly en-route for India but decided to stop-off in Paris where he spent the entire cash float entrusted to him for fuel and other contingencies. He then wired Dalton in Lahore for replacement funds that he also ran through, this time in Tripoli, before returning to England, leaving one understandably very displeased Dalton in his wake.

"The Spider" in India on CD Barnard's 1933-34 display tour.

Part of the Indian team with Captain Dalton and parachutist Harry Ward (extreme right).

The British Hospitals Air Pageants Ltd, founded in 1933 by Jimmy McEwan King and Henry ACG Barker, attracted to its ranks a host of nationally acclaimed pilots operating their own aircraft for a tour of Britain. Headed by Charles WA Scott, who had broken the England-to-Australia record the previous year with a flying time of eight days and twenty hours, the team of famous pilots included

The British Hospitals Air Pageant included many well-known pilots, and seen here are [left to right] WF Drury, Percy Phillips, RH McIntosh, JK Morton, EW Jordan, Pauline Gower, CWA Scott, EW Bonar and Bill Rollason with Fox Moth G-ACCF in the background.

RH 'All Weather' McIntosh, John Pugh, Earl Fielden, WA Rollason, 'Jock' Bonar, Leslie Anderson, EW Jordan, Tommy Nash, Pauline Gower, JK Morton, WAC Kingham, Percy Phillips and Mildred Mary Petre, the latter better known as the Hon. Mrs Victor Bruce.

This intrepid lady, already famous for record attempts on land, sea and in the air, had previously made an unsuccessful attempt to win the women's endurance prize using air-to-air refuelling. Now, invited to join the 'Hospitals' team, she purchased, first the Miles Satyr G-ABVG for aerobatic work and then an ex-RAF Fairey Fox G-ACAS from a Chiswick scrap yard for twelve pounds and ten shillings (£12.50). At McEwan King's suggestion, Mrs Bruce developed the 'Fox Dive' for more adventurous passengers. This involved pulling out of a vertical descent at some fifty feet and flying low over the crowd. With a constant queue of willing participants heeding the announcer Colonel Fitzmaurice's call to "Roll up for the 'Fox Dive' – have the thrill of your life for only fifteen shillings (75p)" this event invariably proved a contender for 'top of the bill'.

But on 14th July 1933, the Fox came to an ignominious end after it caught fire in the air. The pilot on that occasion, John Pugh, managed to put it down at Ford aerodrome near Littlehampton where, fortunately, he was able to rescue his two women passengers from the overturned aircraft before the petrol tank exploded. Upon being handed a note informing him of the accident, the undaunted Fitzmaurice went on to call "Roll up for the Autogiro – have the thrill of your life for only fifteen shillings". Just another day at the office it would seem!

Not so fortunate some six weeks later was the show's parachutist, 'Windy' Evans. A practised performer who had learned his trade as a parachute packer in the RAF, Evans was a determined 'jumper' who would brave the strongest winds. But at Harrogate on 27th August 1933, whilst attempting to counter a crosswind, the canopy of his Russell Lobe parachute collapsed causing him to plummet to the ground with fatal results.

It had been publicly announced that the tour, which competed directly with that of Sir Alan Cobham (see later chapters) would raise some £20,000 for hospitals in the towns visited. In the event the final figure donated amounted to only £6,854. This became the subject of much acerbic comment not only from Cobham, who considered Barker's earlier defection from his own staff a personal betrayal, but from CG Grey then the influential editor of *"The Aeroplane"*. He made the comment rather pointedly that, whereas Cobham was in the business of making people air-minded and earning a little money thereby, Barker was more concerned with exploiting the gullibility of the charity-minded!

Hindsight does suggest that some covering of the financial tracks took place concerning the succession of aerial enterprises run by Barker and McEwan King. Following on from the efforts of the British Hospital Air Pageants came the Sky Devils Air Circus, also owned and operated by this colourful combination's Air Pageants Ltd. Sponsored by Lord Nuffield's Morris vehicle organisation, it commenced its touring season on 15th April 1934 at Stag Lane having announced that it would also donate 10% of its takings to local hospitals. A 'mixed bag' of aircraft comprising three Avro 504Ns, including G-ACLV and G-AECR, an Avro Tutor, CWA Scott's DH.60G Moth G-ACPD, a DH.83 Fox Moth, a Blackburn Lincock flown by 'Jock' Bonar and a Miles Hawk three-seater G-ACPD. These were led by Air Tours' Captain Fielden flying an Armstrong Whitworth Argosy G-EBLF, chartered from Imperial Airways.

Whether Barker and McEwan King's business tactics were truly imaginative or downright evasive is open to interpretation for it is not recorded that any hospitals did actually receive benefit.

The Hon Mrs Victor Bruce carried out the chilling 'Fox Dive' in her Fairey Fox G-ACAS.

Pauline Gower in the cockpit of the Planette which had been decorated by Heath Robinson as a Chinese Dragon for the British Hospitals Air Pageant.

A typical English summer day's flying display with the BHAP in 1933. Aircraft seen are Avro 504K G-EBYW [nearest camera], Spartan Three-seaters G-ABYN and G-ABKK, Fox Moths G-ABUP and G-ACCF, Dragon G-ACCR, DH.60 Moth G-ABOG, Fairey Fox G-ACAS and Monospar G-ABUZ.

The North British Aviation Company was a regular operator between 1929 and 1935. Note the mis-spelling of the parachutist Benno de Greeuw!

Benno de Greeuw was the leading parachutist for North British Aviation.

However, in the following year their seemingly successful, albeit short-lived operations became Jubilee Air Displays Ltd led by the famous racing and long-distance record breaking pilot Owen Cathcart-Jones. His team included such well-known flying personalities as 'Tubby' Hearn, 'Rich' Robinson and Messrs Carleton Ross, 'Bats' Bulmore, Anderson and Steel. It was run in conjunction with the English and Scottish Joint Co-operative Societies which guaranteed that, on proof of purchase of the Co-op's tea, entrance to the flying ground was free. This particular tour was sadly marred by the deaths of the display's director Leslie Anderson and his passenger AJR Jope, when Avro 504K G-ADBS crashed at Bodmin on 15th August 1935.

In 1936, in what proved to be Barker and McEwan King's penultimate promotional venture, 'British Empire Air Displays' was formed although it appears simply to have been a new name and was still formally operated by Jubilee Air Displays Ltd. By this time, the activities attracted less publicity and fewer tour aeroplanes.

The North British Aviation Company Ltd, based at Hooton Park in Cheshire, was originally started up on 2 March 1929 by two ex-Berkshire Aviation Tours pilots, Edward E Fresson and Lancelot J Rimmer. Between 1929 and 1935 it undertook regular tours of the British Isles, initially operating various Avro 504Ks but later including a DH.60G Moth, G-AAWO and a Short Scion, G-ACJI, for pleasure flights and a Wolf Hartz glider flown by Captain Rimmer. Benno de Greeuw, who had once understudied for Douglas Fairbanks Jr in a flying film and Mr N Burt provided the parachute and 'wing-walking' entertainment in a programme that featured the, by now, almost obligatory aerobatic, comedy and stunt flying routines.

Ted Fresson had a chequered flying career either side of his five-year spell with North British Aviation. Like so many other ex-First World War military pilots, he had first sought a peacetime living by trying his hand in the joy-riding game, joining up for a short time with the Cambridgeshire School of Flying. In October 1919 he left for China, where he had earlier lived for several years, in order to design and build a fighter-bomber that was later produced in quantity for the Governor of Shansi Province. However, uncertainties surrounding the political upheaval in the Far East and the death from typhus of his first wife in 1927, brought him back to England and to a subsequent flying appointment with Berkshire Aviation Co.

The time spent touring with both 'Berkshire' and 'North British' offered him his first contact with the highlands and islands of Scotland. Here he recognised the need for regular scheduled passenger, mail and newspaper services and in 1933 he formed Highland Airways Ltd. This was taken over in 1935 by the Whitehall Securities group to become part of United Airways. Fresson was by no means the only provider of joy-riding services north of the Border, for the Scottish Motor Traction Company (SMT), though primarily concerned with bus company operations also extended its scope to include air-taxi and pleasure flights, operating distinctive blue and silver DH.83 Fox Moths from landing grounds throughout Scotland.

Company rivalry was in fact running high in Scotland in the early 1930s. Another concern, Midland and Scottish Air Ferries founded by John and Christine Sword was, by mid 1934, operating a fleet of seventeen aircraft that consisted mainly of DH.83 Fox Moths and DH.84 Dragons. Although competitive joy-riding took place from many sites, including the beach at Ayr, the initial attraction of 'flip' flights later gave way to the more serious business of establishing regular air routes in both Scotland and England. Midland and Scottish was also keen to expand its activities into Ireland with a proposed service between Cork, Dublin and Liverpool. Accordingly, to promote this enterprise, it was arranged that flying displays would take place in both Irish cities.

But a shadow fell over the venture when, on 3rd August 1933, during a rehearsal two days before the first show in Dublin, a mid-

One of the longest company titles in the business, The Portsmouth, Southsea and Isle of Wight Aviation Company Limited undertook both joy-riding and internal airline operations. Seen here are their Klemm L.27A G-ABJX, Puss Moth G-ABIY and Fox Moth G-ACIG.

air collision between two Avro Cadets belonging to the Irish Air Corps resulted in the death of Lt. Jim Twohig. More tragedy occurred on the day of the show itself, when, during a mock attack by three fighters, an Irish Air Corps Vickers Vespa observation machine was caused to spin into the ground. Both occupants, Captain Oscar Heron and Private Richard Tobin later died from injuries received in the crash. Despite these unfortunate happenings, a fleet of twelve visiting aircraft entertained a crowd of some 30,000 in the Phoenix Park with joy-riding provided by Midland & Scottish's Airspeed Ferries, G-ACBT and 'FB and Avro Ten, G-ACGF. Because of the fatal events, the Irish Air Corps withdrew from the Cork Air Carnival held at Ballincollig over the following two days.

Though undertaken to advertise the airline's new service, which opened on 14th August 1933, poor passenger numbers and a government insistence that Midland and Scottish form an Irish company, saw the operation cease on 30th September. The company itself closed down the following year.

At this time, Britain was seemingly awash with airshows, not only those performed by the larger touring fleets as described, but also with the kind performed by companies such as Modern Airways Ltd which, trading as The Crimson Fleet, essentially flew from fixed bases, in this case Knotts Farm, Kingsdown near Wrotham in Kent and also later, Porthcawl. RJ Bunning, based in Pontypool, was another who, in the late 1920s, flew an Avro 504K, G-EBSM, in an attempt to re-vitalise the flagging South Wales pleasure flight trade But, like so many, he met with only limited success.

British Flying and Motor Services Ltd was another which, formed in May 1928, fell into the fixed-base category. Before being taken over by Inland Flying Services in 1929, it used two DH.6s G-EBPN and G-EBVS, a DH.60 and an Avro 548, G-EBPJ to carry joy-riders from Maylands Farm at Romford in Essex. As the decade drew to a close, Inland Flying Services went on to work the Essex fields with Avro 504Ks, G-AAFE and 'FT before turning its attention to the Isle of Wight where, at Apse Manor, Shanklin, in February 1930, it was renamed Wight Aviation Limited.

By 1931, both Avros had been retired and replaced by a three-seater Klemm, G-ABJX, a Spartan Three-seater G-ABLJ and DH.60 Moth G-AAAG. In 1932, the company was again retitled

and became Portsmouth, Southsea and Isle of Wight Aviation Ltd, primarily to undertake air ferry operations between Portsmouth, Bournemouth and Southampton to the Island. However, joy-riding activities still continued to play a significant part in the company's business with its new senior pilot, Charles Eckersley-Maslin, frequently using director Lionel Balfour's Puss Moth G-ABIY for this purpose at Portsmouth or Apse.

Major EG Clerk, having founded the Lancashire School of Aviation in 1928 with three Avro 548s, G-EBIT, 'JU and 'OK, also went on to become a 'fixed based' operator. In 1934, his Kinmel Bay Flying Services flew Rhyl holidaymakers from the shore in Avro 504Ks G-EBYW and G-ABWK. He then took these aircraft, along with a Spartan three-seater, G-ABJS as a 'flying circus' to Toowoomba in Australia for the winter season. The following year, he returned to England and, with Avro 504N, G-ACRS, operated from Rye on the Sussex Coast as Alpha Flying Service.

Two extremely popular figures who appeared on the largely male-dominated display circuit in the early 1930s were Pauline Gower and Dorothy Spicer. Their first operations involved taking up passers-by in a hired DH.60G Gipsy Moth, G-AABK from a field near Sevenoaks in Kent. With Pauline concentrating on the flying and Dorothy attending to engineering and maintenance matters, their first joy-riding endeavours proved moderately successful. Encouraged by this, they took delivery of a two-seater Simmonds Spartan, G-AAGO. In August 1931, under the name Air Trips, they were, as recorded in *"The Aeroplane"*, carrying passengers from Mr John Frost's Oakley Wood aerodrome near Wallingford in Berkshire and several other locations, including Cowes, Reading, Stag Lane, Heston, Northolt and Didcot. Here, on one occasion, a concerned Miss Gower had to hit a drunken would-be 'wing-walker' on the head with a spanner to prevent him leaving the cockpit while in flight.

In May 1932, they enthusiastically accepted an invitation to join Modern Airway's Crimson Fleet air circus. This introduction to the touring life soon highlighted the pressures placed upon all airshow performers. Within a five-week period and now operating a three-seater Spartan, G-ABKK *"Helen of Troy"*, so-named, they explained, because 'it was a Spartan and sometimes goes wrong!' Miss Gower gave 1,065 joy-rides and six displays of formation

Lancashire School of Aviation's Avro 504K G-EBIT, seen here with senior pilot, ED Ayre.

flying and aerobatics. In the course of this she experienced two crashes and one engine failure. It is perhaps surprising that more mishaps did not occur for, as she later observed, it was common practice when coming in to land for passengers to stand up in the front cockpit waving caps and handkerchiefs to admiring friends and relatives on the ground, but which obscured the wretched pilot's view. Fortunately the Spartan's landing speed was low and the airframe reasonably forgiving when contact with the ground proved heavier than intended.

After spending the winter of 1932-1933 recuperating and reflecting upon their business future, the partners came to the conclusion that the only way to reduce a growing overdraft would be to continue on the airshow circuit for another season. This time

Pauline Gower and Dorothy Spicer seen here with their Spartan Three-seater G-ABKK "Helen of Troy" at Stag Lane in March 1932.

it was the British Hospitals Air Pageant and the persuasive patter of promoter Jimmy McEwan King that secured their services. Pauline Gower now found herself flying her Spartan three-seater alongside a similar machine, G-ACAF belonging to Percy Phillips. She later commented that it was this organisation, with its fifteen renowned pilots and three parachutists, which provided an experience and generated a confidence that would have been impossible to acquire in any other way.

One of the locations visited by the BHAP had been Hunstanton on the Norfolk coast. Making full use of contacts made during that visit, the ladies, now operating as Air Trips Limited, decided to set up a camp of their own at this resort's South Beach in 1934 and 1935. A novel feature that proved highly popular with holidaymakers was the arrangement made with Geoffrey Searle and Jack Tibbetts, the local owners of a speedboat, *"Tigress"*, that allowed Miss Gower's passengers to flour-bomb those in the boat below. Though undoubtedly great sport for all concerned, the combination of flour and water was, one suspects, guaranteed to provide some pasty complexions!

Bruce ('Bill') Williams, a professional actor and part-time stuntman whose parachute bore the message 'I'm Falling For You', was also taken on and was encouraged to learn to fly a Klemm machine purchased by the speedboat entrepreneur. However, this particular episode came to a rapid end when Williams wrote the aircraft off in a taxiing mishap before it was insured. With little competition, the sea-side business proved to be brisk – no doubt being encouraged by the company's slogan 'Stop me and fly one' featured on top of its van. It was therefore decided to purchase a DH.83 Fox Moth, G-ADNF with extra seating capacity. But on 11th August 1935, the day following its arrival, disaster struck when engine failure on the first take-off resulted in an ungainly descent into swampy ground. The engine was torn from its mountings and the two passengers in the enclosed cabin, Mr Roy Ball and another gentleman who was reportedly unnamed 'because of his wife's ill-health', were hurled forward through the resulting hole. Amazingly there were no serious injuries and in due course the new acquisition was repaired and sold on through various owners to end its days as VH-ABQ in the gold fields of New Guinea.

The ladies' forays with the air circus tours of 1932 and 1933 were not to be their last. In 1936, as part of Tom Campbell Black's British Empire Air Display, Pauline Gower, now the most qualifed woman pilot in the world, became its chief pilot. Here, she flew alongside not only Campbell Black himself but other talented colleagues such as Eric 'Jock' Bonar, who had taught Jim Mollison to fly, (Hawker Tomtit, G-AEES), Robert 'Dotty' Doig ('Flying Fleas', G-ADPW, G-AEEW), Cressy Reynolds (DH.83 Fox Moth, G-ACKZ), and 'Bogie' Grey, 'Gerry' Chambers and Herbert B G Micklemore (Avro 504Ns, G-AECR, G-ACZC and G-ADEI). Dorothy Spicer, then the only woman in the world to possess the Air Ministry's A, B, C and D Ground Engineer's Licences was appointed the Display's Senior Engineer.

The 1936 tour got off to a bad start however, when, at Hereford on 1st May, Walter Cadic, a naturally-gifted twenty-six year old pilot engaged to fly the display's Short Scion, G-ADDT, took up a visiting BAC Drone powered glider, G-AEBC. A spin, entered at some 400 ft, left no margin for recovery and the resulting crash proved fatal. Harry Ward, one of the show's three parachutists, had earlier had a short flight in the Drone, during which his seat had disconcertingly slipped back. With the rudder bar out of reach he had found it difficult to control the machine and Ward was left in little doubt that it was a repeat of his experience that brought about Cadic's downfall. Unsurprisingly, the death of this popular young man, who had gained his 'wings' at the Gravesend School of Flying just two years previously, cast a deep pall over the tour proceedings.

Shortly afterwards at Coventry, on 10th May, extremely gusty wind conditions contributed to the five-seater Scion, flown by Micklemore, failing to clear a hedge on take off. The loss of a wheel led to a landing, that, while incurring further damage to the

Bruce 'I'm Falling For You' Williams at Hunstanton with Air Trips Ltd.

aircraft, fortunately allowed all on board to emerge unhurt. More dramatic was the accident that occurred later the same day, though again, miraculously, no lives were lost. Pauline Gower recounted how, having commenced her take-off run in a demonstration air race, she collided with 'Bogie' Grey's machine attempting to become airborne from a different direction. Both Miss Gower's Spartan, G-ABKK and Grey's Avro 504N, G-AECR were totally written off and Miss Gower spent three weeks as the unwilling guest of the Coventry and Warwickshire Hospital. Fortunately, however, she was able to re-join the tour with a replacement Spartan, G-ACAD at Newcastle on 4th June.

Tom Campbell Black was, at the time, a name famous throughout the country, as indeed was that of his wife, the actress Florence Desmond. An ex-East African farmer and bush-pilot, he had undertaken commercial flying work for Wilson Airways and later Lord Furness, before teaming up with CWA Scott to win the 1934 MacRobertson Air Race from Mildenhall to Melbourne, in DH.88 Comet, *"Grosvenor House"*. But, on 19th September 1936, whilst at Liverpool's Speke airport preparing for the Schlesinger air race from Portsmouth to Johannesburg, he met a distressing end. His parked Percival Mew Gull, G-AEKL was run into by an RAF Hawker Hart, K3044, its propeller slicing through the cockpit and the unfortunate occupant.

Reverting to Gower and Spicer, although Air Trips Ltd moved on to Hayling Island after Campbell Black's tour finished and began the 1937 season with pleasure flying from its new base, the company finally closed its doors soon afterwards. Both ladies now wished to pursue more stable interests although both went on to enjoy highly successful careers in wartime aviation. But, on 23rd December 1946, Dorothy Spicer, now Mrs Richard Pearse following her marriage in April 1938 to the founder of aircharter business Wrightson & Pearse, was killed along with her husband when the Avro York, LV-XIG, in which they were travelling to start a new life in Brazil, crashed into a mountainside near Rio de Janeiro. Ironically, just a few weeks later, on 2nd March 1947, Pauline Gower, now married to Wing Commander Bill Fahie, also died in tragic circumstances after giving birth to twin boys.

The loss of two such kindred spirits, still only in their mid-thirties, was deeply felt throughout the British flying community, which felt they undoubtedly still had so much more to give.

Hawker Tomtit G-AEES was flown by Jock Bonar on the British Empire Air Display 1936 tour.

The wreckage of Pauline Gower's Spartan G-ABKK and Bogie Grey's Avro 504N G-AECR at Westwood Heath, Coventry on 10th May 1936.

Harry Ward prepares his 'bird-man' wings for a delayed drop at Sherburn-in-Elmet in August 1936.

Following Campbell Black's death, promoters King and Barker decided to form 'The Aircraft Demonstration Company' to trade as the 'Coronation Air Display' in 1937 for what was to be their last tour. Strongly challenging the flying circus tours were the Royal Air Force's 'Empire Air Day' displays which had now become an annual event. Highly professional, they provided an ideal opportunity to demonstrate the latest military aircraft, and the polished aerobatic skills and daring of Britain's Service pilots. In 1937, no less than fifty-three RAF stations and civil airports staged such shows throughout the country.

This level of state-supported competition clearly diluted the attractions offered by McEwan King and Barker. After performances in thirty-two towns on the British mainland and eight in Ireland promoted by the Irish Aero Club, the level of takings was judged not to be worth the effort and with debts accruing at an alarming rate, this final tour simply ground to a halt. When, despite an impressive display of financial gymnastics Barker and King appeared before the Official Receiver later in the year, they were found to have liabilities of £6,000 and assets amounting to no more than one aircraft valued at £177, and ten shillings (50p) cash-in-hand.

Operating as they all did to a very similar format, Britain's flying circuses had placed a great reliance on the same relatively small band of display and joy-riding pilots, 'wing-walkers' and parachutists – the latter often considered by the public to be the most daring of all the performers. Sir Alan Cobham likened the crowd's tense anticipation that preceded a jump to what one would associate with an ancient gladiatorial contest or a modern bullfight. It was therefore, entirely by design that these brave individuals, both men and women, were always called upon by the show organisers to provide the last act of a display. As happened all too frequently, a misjudged descent which resulted in injury or worse would always send a crowd home feeling it had had its money's worth!

After 1925 when it was belatedly decided to equip Service aircrew with parachutes, a number of 'jumpers' had become demonstrators for the various equipment suppliers, or joined the newly formed RAF display team based at Northolt. One such character was Harry Ward. After touring RAF stations giving 'pull off' and 'free fall' displays, he had left the Service in 1929 to become, more prosaically, the driver of a London bus. But he was not destined to remain on the ground for very long. Having co-founded what was commonly known as the London Bus Drivers Flying Club, he marked its opening in 1931 by making an exhibition jump in plus-four trousers and a driver's hat. He went on to become a principal attraction with Sir Alan Cobham's organisation in 1933, India Air Pageants in 1934 and the succession of touring displays run by Barker and King.

It was Tom Campbell Black who persuaded Ward to introduce the 'birdman' stunt into his list of flying circus aerial events. First performed in America by Clem Sohn before he brought it over to Britain in 1936, this spectacular event required a parachutist to equip himself with a wooden framework over which fabric was stretched to simulate bat-like wings and tail. The theory was fine – a 'jumper' simply launched himself into space and extended his arms and legs to execute swooping bird-like manoeuvres, before deploying his parachute for a safe landing.

Practice, however, often brought serious problems. In 1937, Sohn, after exercising his skills before British crowds, then visited France, where his first performance before an estimated crowd of 100,000 at Vincennes on 25th April, turned out to be his last. Jumping from 10,000 feet, Sohn glided with wings extended as planned, before deploying his main parachute with some 1,000 feet remaining. His main parachute deployed but failed to open fully, whereupon Sohn immediately pulled the rip-cord of his reserve canopy. But this too refused to function correctly and with both his lifelines entangled, the American plummeted to the ground at some 100 mph. Although Ward volunteered to complete the remainder of the French tour, Sohn's manager turned the offer down, with the rejoinder 'one dead birdman is quite enough'.

Ward's own role-model was the Danish-born John Tranum, an aerial performer who had spent most of his early life 'barnstorming' his way round America as pilot, 'wing-walker' and parachutist. A fearless yet intelligent approach to his art had quickly marked him out for attention by exhibition promoters. They recognised that if there was a new angle to be explored, he was the man to do it. He proved this beyond doubt when he rode a motorcycle over a cliff, successfully descending by parachute, when many other attempts had resulted in serious injury or death.

But even the most skilful parachutists occasionally came unstuck as Tranum learned to his cost and the amusement of others at a flying meeting held at Brooklands early in 1930. He had agreed to perform only on condition that he be presented to each of the chorus girls from the London musical show 'Silver Wings', attending the display. But with his mind probably on the social rewards likely to follow, he badly misjudged the strong wind and landed, not as planned with professional precision in front of a bevy of admiring ladies, but firmly in the middle of the infamous sewage farm on the airfield boundary. His introduction to the delights of show business did not take place that day!

In September 1931, Tranum in partnership with long-distance flyer Oscar Garden, founded Skywork Ltd with the aim of taking 'The Spartan Air Circus' to South Africa. Equipped with three Hermes-powered Spartan three-seaters, G-ABPZ, G-ABRA and G-ABRB and one Desoutter G-AAPP, they left Britain on 9th October with a small team of pilots that, in addition to Garden, included JR King, CEF Reilly, ED Ayre and ED Cummings, largely recruited from CD Barnard's organisation. Displays were given at thirty-two locations in the Union between October 1931 and February 1932 and at the outset the tour proved a phenomenal success. Britain had just gone off the gold standard and as local joy-riding and entrance fees were being paid for in sovereigns, a colossal profit appeared to be there for the taking.

At Port Elizabeth however, a woman who poked her finger into a spinning propeller, sued for damages and when judgement went against the touring management, all assets were seized and a much-chastened team returned home. Skywork was immediately disbanded on its return, having sold one of the Spartans G-ABPZ to offset the damages before leaving the Union. Garden later returned to East Africa with a Spartan where he continued joy-riding for several years, sponsored by the Vacuum Oil Company (Mobiloil) before returning to Britain as an airline pilot in 1935.

Tranum returned to Denmark and resumed his freelance jumping activities but inevitably perhaps, he pitted his skills against stacked odds once too often. On 7th March 1935, whilst in an aircraft that was climbing to reach 33,000ft, an altitude from which he hoped to regain the world record for a delayed descent, he signalled that he was experiencing trouble. The pilot landed as quickly as he could, but was shocked to find Tranum dead in the cockpit, having apparently succumbed to a heart attack.

Harry Ward, along with ex-show jumpers Bruce Williams and Bill Hire, later became a leading instructor at the Royal Air Force's parachute training school at Ringway during the early years of World War II. Blessed with a charmed life, he continued to fly whenever he could and died in 2000 at the great age of ninety-seven.

Doubtless, the touring air shows had engaged and indeed created, a wide variety of artistes and organisers who possessed many admirable talents. But all too often, the shows were bedevilled by lack of capital and a poor understanding by their promoters of the special needs and exacting requirements that attached to a touring display. One man however, with an outstanding flair for organisation, stood head and shoulders above the rest. His exhibitions set the standards and provided a classic example that others tried to follow. His name was Sir Alan Cobham.

The DH.61 Giant Moth G-AAEV, named "Youth of Britain" after purchase by Cobham for his Municipal Aerodrome Campaign.

Chapter Four:
Enter Cobham – 'Ringmaster Extraordinaire'

'Before you pay your fare to fly,
And leave the ground to soar on high,
Think of your children and your wife
Be prepared – insure your life.'

A verse which, after a flight taken by the company's advertising representative,
was amended the following day to read:

'To pay your fare and fly by air,
Is safer far than tram or car,
Or walk in street, or ride by bus,
But safer still, insure with us.'

The London Guarantee and Accident Co Ltd advertisement at the
'Huddersfield Flying Exhibition' December 1920

Such then was the chequered background against which Sir Alan Cobham, the Berkshire Aviation Company pioneer, and already a household name following his well publicised Empire trail-blazing flights in the 1920s, decided to make the British public and government of the day seriously 'airminded'. Convinced, by the success of his early pioneering endeavours, that it would be the aeroplane and not, as was then thought in high places, the airship, that would provide the next major step forward in transportation, he sought every available means to broadcast his message.

In 1926, reasoning that trains could not run without tracks, ships without harbours or cars without roads, he had entered into correspondence with every major municipality in the British Isles to plead his case for a network of first-class aerodromes across the country. However, though his efforts were regarded in many official quarters with genuine interest, little positive action resulted.

It was in 1929 that Cobham became a founding director of National Flying Services Ltd, an organisation formed to set-up and co-ordinate the activities of the increasing number of flying clubs around the country. This new position also assisted his self-appointed role as arch propagandist for British aviation and having again received the backing of Castrol oil magnate Lord Wakefield, he set about organising a highly ambitious aerial tour of over one hundred towns and cities. His aim was simple and direct – to impress upon all the Lord Mayors, mayors, councillors, aldermen, town clerks and borough surveyors that, unless they moved with the tide of progress in aviation and recognised the need for a local aerodrome, their towns would be doomed to languish in the backwaters.

He purchased a ten-seat DH.61 Giant Moth airliner G-AAEV from de Havillands which he christened *"Youth of Britain"*, engaged Dallas Eskell, then traffic manager at Imperial Airways, to become his general manager and hired two specialist ground engineers. Then, Cobham set off on an exhausting twenty-one week campaign during which he visited 110 towns and carried, in 5,000 flights some 50,000 passengers including 10,000 schoolchildren paid for by an 'anonymous benefactor' (Lord Wakefield).

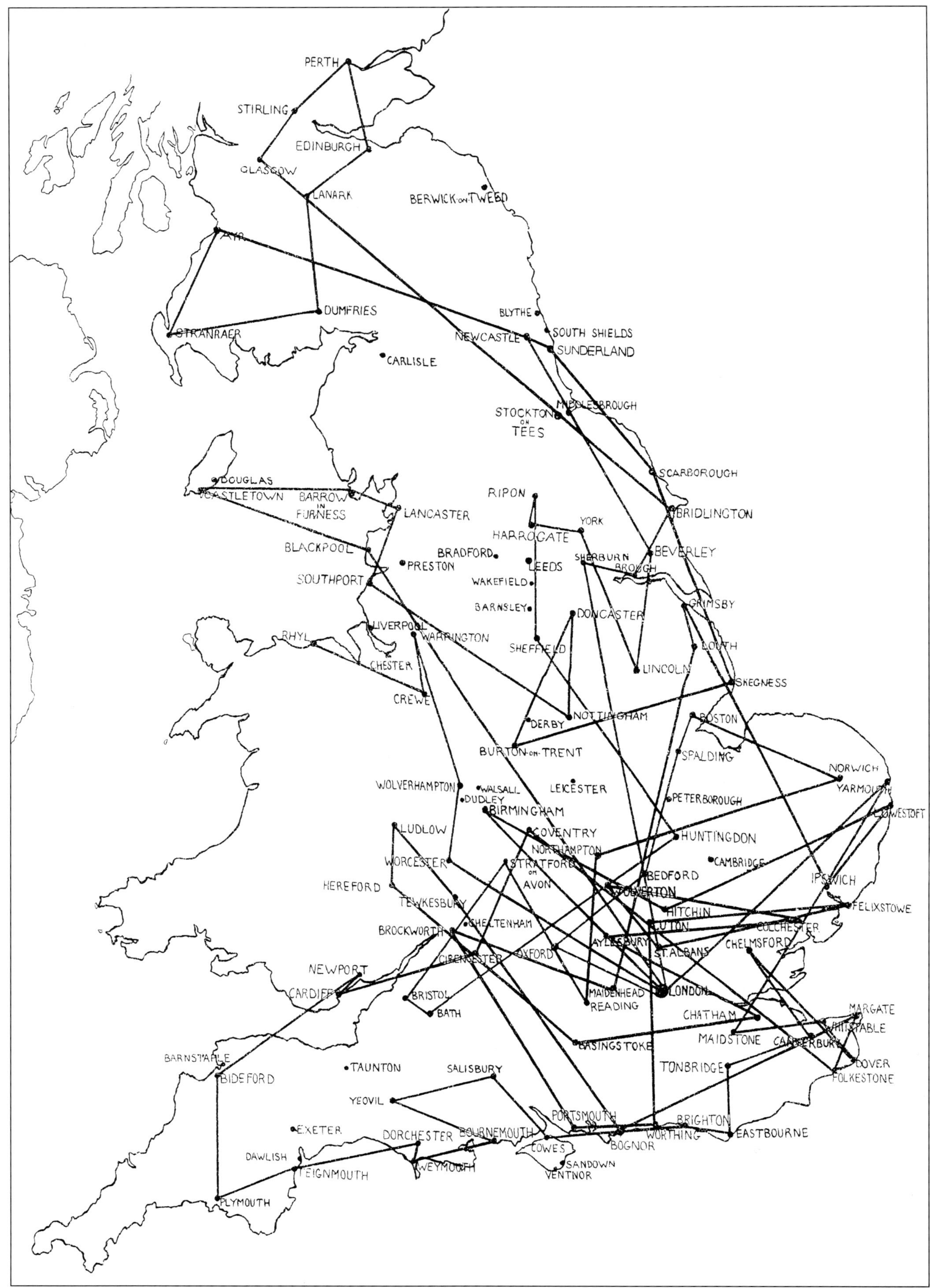

The area covered by Cobham in his 1929 tour of Britain.

Sir Charles Wakefield was a philanthropist and the anonymous donor who supported many aviation causes and donated 10,000 free flights for schoolchildren in the "Youth of Britain".

Sir Alan Cobham who, in 1929, determined to undertake a Municipal Aerodrome Campaign and to make the public 'airminded'.

The daily routine varied little. Weather permitting, he arrived at his destination at eleven o'clock each morning and following the welcoming ceremony took the civic dignitaries aloft. After this, local schoolchildren were given the undoubted thrill of seeing their homes from the air. Discussions over lunch usually provided further opportunities for Cobham to 'spread the word' as did his meeting with the general public during the pleasure flights he operated throughout the remainder of the day. Cobham later judged his mammoth effort very successful, having experienced only occasional minor incidents that temporarily prevented flying.

A business-man as well as an enthusiast, Cobham subsequently received sizeable commissions from several municipalities for aerodrome site surveys, but on at least one occasion his enthusiasm and opportunism appear to have preceded an official go-ahead. According to a report contained in a Yorkshire newspaper, the *"Bridlington Chronicle"*, local councillors expressed surprise at receiving a bill for £82.11s.3d (£82.57) from Alan Cobham Aviation, for what they regarded as an unauthorised report relating to possible sites for a local aerodrome. The account went on to say that, although undertaken following a misunderstanding, Cobham had undoubtedly gone about the work in good faith. As men of honour, however, the council intended to pay up and indeed, most commendably, its members did not even intend to look at the report until they had done so!

To set the seal on this 'Round Britain' venture, a large conference that included some 250 representatives of local governing bodies was held on 5th November 1929 at the Institute of Electrical Engineers in London. It was followed the same evening by an equally impressive dinner at the Savoy Hotel. This expensive effort was aimed at getting the public and, let it not be overlooked, Cobham's own commercial future, off the ground.

But this particular vision was never to be realised on the scale that he had hoped for. Many other events, not least another world war, would direct the course of air transportation before it could be fairly claimed that the aeroplane had significantly influenced the general way of life in Britain.

The future of aviation then still uncertain, Cobham decided in 1931 that a much greater impact had to be made on the national mentality. As he said, "I was cudgelling my brain for something substantial that would embed itself in the public consciousness as deeply as Pancake Tuesday or Fireworks Night" and it was this approach that encouraged his conviction that every city and town should have its own annual Aviation Day. He therefore set about turning what he dismissively regarded as 'casual touring' by other organisations into a national institution. Cobham's flair for organisation was matched equally by his prodigious energy and motivation. Through the ensuing months, assisted again by his Municipal Aerodrome Campaign colleague, Dallas Eskell, he hammered out the requirements and aims not only for a summer tour of the United Kingdom in 1932, but also for a tour of South Africa in the following winter of 1932/33.

Cobham's immediate key objective was to gather one million signatures in the British Isles to endorse what he called his 'Mandate'. This was a document (see Appendix I) in which the following organisations were each invited to contribute appropriate words of support:

> The Royal Aeronautical Society
> The Royal Aero Club
> The Air League of the British Empire
> The General Council of Associated Light Aeroplane Clubs
> The Guild of Air Pilots and Air Navigators
> The British Gliding Association
> The Association of British Chambers of Commerce
> The London Chamber of Commerce – Civil Aviation Section
> The Automobile Association
> The Society of British Aircraft Constructors
> The Society of Model Aeronautical Engineers
> The Model Aircraft Club

Cobham thereby hoped to influence the government to take a more urgent and energetic attitude towards the development of civil aviation throughout Great Britain.

Thus, with his broad aims now established, he began to assess his needs for machines and personnel. Small aerobatic aircraft were plentiful, but he realised that a highly specialised type of machine would also be essential, one that could carry, say, ten people in and out of small primitive landing grounds, provide excellent views unobstructed by wings or wheels, and would allow passengers to enter and leave rapidly. "After all", he reasoned, "time spent on the ground was money wasted." Such an aircraft, however, was not readily available.

Cobham therefore presented a specification and an order for two small multi-engined airliners, priced at £5,195 each, to the fledgling Airspeed Company, then occupying a disused bus garage in York. This was a company of which, following persuasion from ex-de Havilland colleagues, Hessell Tiltman and Neville Shute Norway, Cobham had become a founding director in March 1931. His act of faith paid off. By the time his diverse collection of aeroplanes eventually gathered to launch the tour at Hanworth on 12th April 1932, the first Airspeed AS.4 Ferry G-ABSI, powered by three 120hp DH Gipsy III engines had undertaken its maiden flight at Sherburn-in-Elmet and received its Certificate of Airworthiness following trials at Martlesham Heath. It joined what Cobham now called the National Aviation Day fleet just three weeks later on 4th May. The second machine, G-ABSJ arrived on the tour shortly afterwards in June.

The new airliners, christened *"Youth of Britain II"* and *"Youth of Britain III"*, in the capable hands of H C Johnson, J D Parkinson and Joe King carried some 92,000 passengers in the 1932 season

Illustrative of the business side of Cobham's tour and his discussions with local dignitaries; here he is with Hull's Luncheon Club President, Tom Shepperd, at Brough.

The Airspeed Ferry, designed to Cobham's specifications, had a short field performance that served it well during the National Aviation Day tours of 1932-35.

Ex-Imperial Airways Handley Page W.10 G-EBMR frequently led the Cobham touring formations.

alone, and went on to become the workhorses of the tours that took place throughout Britain during the next four years. Cobham also eventually obtained two larger Handley Page W.10 airliners, G-EBMM *"City of Melbourne"* and G-EBMR *"City of Pretoria"* from Imperial Airways, the intention being that 'MR should serve as a reserve machine. Conflicting performance capabilities meant that whereas the Ferries could become airborne with a full passenger complement in take-off distances of less than 200 yards, the much larger Handley Page machines could only be called upon when longer grass strips were available.

In order to operate the varied types of demonstration and passenger-carrying aircraft now forming his Display, Cobham had also to find flying and support personnel who not only possessed the appropriate qualifications and experience, but additionally, were willing to undertake what promised to be a nomadic life-style for long periods of time. To assist in his task he contacted Captain AG

Lamplugh, senior assessor for the British Aviation Insurance Group and whose knowledge regarding pilots with the best safety records was second to none. One of his first recommendations was to employ Hugh Johnson who immediately made Cobham aware that his previous employers, the Ford Motor Company, had spare hangar accommodation at its airfield in Sussex. It was following this suggestion that Ford Junction aerodrome, named after the nearby village and not, as often erroneously stated, the Ford Company itself, was chosen to become Cobham's flying and maintenance headquarters.

Johnson's private notes also recall that he was almost immediately called upon to stand in for Sir Alan's piloting role in a film *"The King's Cup"* then being filmed at Hanworth. With several purposely produced bad landings and the odd crash sequence called for, Johnson concluded that this requirement somewhat exceeded his original terms of employment!

The main difficulties facing Cobham and Eskell were the choice of towns and cities they should visit and the securing of local landing sites that would meet the demands of an Air Ministry already, in Cobham's opinion, weighed down with excessive bureaucracy. To prove his point, he retained for many years, a telegram granting qualified approval to use a landing site at Melton Mowbray in May 1932, which read as follows:

"To: National Aviation Day Ltd

Melton Mowbray Aerodrome Licence 2530 hereby extended to cover Handley Page W10 subject to reduction of maximum permissible weight to 11,500lb and Airspeed Ferry subject to reduction of maximum permissible weight to 5,100 lb. No flying with these types when wind between W-by-S and NNW or W-by-West and NE-by-North or between E-by-North, SSE or S-by-East and SW-by-W. 20ft trees in SW corner of site to be felled before flying with these types commences. These limitations have been notified to Police.   Air Ministry"

One has to concede that Cobham's enduring disdain of Ministry bureaucracy was not without justification! It was also necessary to take account of potential clashes with other visiting attractions such as Bertram Mills Circus and, heaven forbid, any other aerial competition likely to appear in the neighbourhood. National and local sponsorship along with general publicity matters required constant attention. Such factors were the cause of frequent disagreements between Cobham, who naturally relied on his previous touring experience, and Eskell, who considered that his exposure to airline practice now afforded Cobham a more 'professional' and up-to-date approach.

The NAD 1932 flying team before the introduction of uniforms, with Dallas Eskell [extreme left] and Sir Alan Cobham [extreme right]. Shown standing are JD Parkinson, CK Turner-Hughes, R Ogden, AH Rawson, unknown, ED Crundall, CW Bebb and HC Johnson and sitting WA Rollason and EB Fielden.

Ford aerodrome near Bognor Regis was the winter maintenance and overhaul base for the NAD aircraft.

Cobham's extensive coverage of the country is evident in the spread of locations for a summer tour.

Despite their differences, perhaps even because of them, organisational weaknesses and possible areas of trouble were soon identified. A communication system was set up that subsequently ensured an efficient three-way daily link between the touring unit, Ford aerodrome and the London office in Trafalgar Square where Edwin Rossiter and Leslie Castlemain had become Cobham's financial and publicity mainstays.

Control of the touring itinerary at times proved difficult, threatened as it frequently was by landowners and farmers who had seemed keen enough to co-operate when first approached, but had subsequently forgotten about their original commitment. On more than one occasion, Cobham's advance guard arrived to find that trees or fences had not been removed as previously agreed and that a proposed landing site had been put to the plough. Such discoveries naturally called for a rapid reassessment of other suitable local areas or more frustratingly, cancellation of the display with the unavoidable adverse effect on publicity and subsequent loss of revenue.

In the early months of 1932, with the pre-tour preparations almost complete, Cobham was beginning to view the increasing number of staff and the attendant wage bill with some concern. Obviously the daily takings, so dependent on good weather, aircraft serviceability and an accident-free reputation, would have to be substantial and consistent. His concerns on this score were, however, balanced by the optimistic thought that his well-planned organisation would be able to cope with any contingency.

Reporting directly to Dallas Eskell, were the chief engineer Mr Hutchinson, ground engineers Allison, Cook, Gribble, Williams and Roy Bonner, the initial chief pilot Capt EB 'Batty' Fielden and Mr Page, the screening foreman, who was responsible for the siting of all enclosures and display equipment. Catering, laundry, stores and the camp staff who looked after the erection of tents and baggage transportation, cleaning personnel and a private police force which patrolled the enclosure boundaries and car parks all swelled the ranks of support staff.

Five or six weeks prior to the show's arrival it was arranged that the London office would inform the appropriate local authorities and newspaper editors of the forthcoming event. At the 'three-weeks-to-go' stage, a first advance of two or three men would descend on nearby towns with instructions to co-ordinate publicity with main stores and places of entertainment. Ten days ahead of the show, a second contingent would arrive to carry out a final check on all the pertinent arrangements and, with seven days remaining, banner staff would turn up to fix streamers and post bills on hoardings etc. The

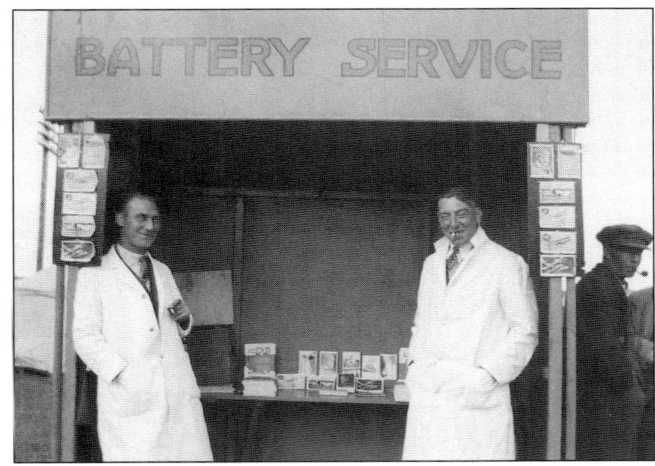

George Ketchell [left] was a sharp-minded London street vendor charged by Cobham with selling tickets, merchandise and programmes.

'night-ahead' man's job was to ensure that all was in order for the display convoy's imminent arrival at the landing field. Finally the 'arrower', often assisted by Cobham's younger son Michael, had the task of attaching arrow signs to lamp-posts and telephone poles pointing out the way to the landing ground. With such detailed planning it might be asked what could possibly go wrong?

The only staff recruited from local sources were the gate men, whose function was to take the entrance money. Often, however, the Corps of Commissionaires could only supply older men who, under pressure from an onrush of people, were quite incapable of separating their own cash from what little genuine entrance money they had managed to take. This expensive oversight was quickly rectified with the appointment of Ron Price as permanent gate manager, and the employment of capable hands more used to exchanges of cash in fast-moving situations. Two assistants were taken on – the Ketchell brothers – who were London street vendors and who soon demonstrated an ability to 'supplement' their official wage with fast talking and the short-changing of an unwary public. On one occasion, George Ketchell was reproved by the Chief Constable of Sheffield Police for having given an incorrect amount of change, whereupon George, completely unabashed, pointing to a few coins he had surreptitiously let fall to the ground, admonished him in turn for his carelessness! Somewhat aggrieved, Hugh Johnson, who had later become chief pilot, remarked to Cobham, that he reckoned the income of both George and Cecil well exceeded that of his own!

Keeping the wide variety of road vehicles serviceable was a major problem.

Sir Henry Segrave's record-breaking "Golden Arrow" was an added attraction on the 1932 NAD tour.

Throughout the touring years, Cobham was greatly irked by the large groups of people who gathered on nearby high ground to obtain a free view of the flying. Caustically, he dubbed them the 'Aberdeen Grandstand' or 'Hedge Guests', and over the years their numbers probably well exceeded those that paid an honest entry fee to see the shows. Today, with hindsight, it is possible to have more than a little sympathy for those who, in the years of the Depression, probably could not remotely afford the entrance fee for themselves, let alone a family. The local excitement which had been fuelled by all the advance publicity must have made the temptation to obtain a free viewing irresistible. It certainly proved so for many small boys who, having cut slits in the six-foot-high canvas screening and stuck their heads through, soon felt the heavy hand of a patrolling policeman in a stinging reprimand.

One of the greatest difficulties facing the display organisation was the lack of standardised transportation. Eight different types of road vehicle, initially sponsored by the Austin Motor Company, had to travel some sixty miles each day and required an extensive range of spares. Fortunately Sir Henry Segrave's world speed record winning car, the *"Golden Arrow"* (now exhibited in the Beaulieu Motor Museum), which provided an added attraction on the tour, was not transported under its own power! The maintenance problem became even more complex when it came to supporting a variety of aircraft that included, in addition to

Flown by AH Rawson on the 1932 NAD tour, the Southern Martlet G-ABBN is seen at Shoreham, where it was built.

Cobham's own DH.9, Gipsy and Tiger Moths, Avro 504s, a Comper Swift, Cierva C.19 Autogiro, Desoutter, Spartan and Southern Martlet machines and a BAC Vll glider in addition to the larger Airspeed Ferry and Handley Page W.10 airliners.

On one occasion, Hugh Johnson recalled that he collected, at Cobham's request, a Blackburn Bluebird from Brough. On rejoining the touring show at Ilford, the chief engineer Hutchinson shook the tailplane and declared the machine was falling to pieces. Nevertheless, Johnson carried on and in his words 'carried as my first passenger a man wearing a bowler hat. I didn't know who was more frightened, him or me!'. Following this disconcerting experience, Hutchinson declared the aircraft unfit to fly, but only after Johnson had flown it, after its short lived appearance, back to Blackburn's factory at Brough.

With so many display aircraft carrying out large numbers of short flights and each having to taxi up to several miles over rough ground each day, the maintenance and servicing requirements were extremely demanding, calling for much improvisation, personal initiative and at times great patience by all when dealing with a boss determined to exert his authority. Percy Allison, engaged at the outset of the 1932 tour as a young ground engineer, was one employee who soon felt the edge of Cobham's tongue when he considered it an unnecessary waste of time to repair cracked exhaust pipes on the Airspeed Ferry.

Following what Allison described as a 'forthright exchange of views' his illustrious leader stormed off and avoided him for several days. Allison later commented that, although nominally fired on more than one occasion by an irate Cobham, he managed, over the next forty years or so to remain on the payroll. He simply kept a low profile until the crisis of the day blew over and the 'governor's' attention became focused on more pressing matters, not the least of which on the first tour was the question of fuel and oil supplies.

For Cobham's earlier long-distance ventures, Shell had produced the petrol and Wakefield (Castrol) the oil, but Shell had now also moved into the lubricating business and naturally wanted to see their oil products overtly advertised. In the event, and

The Grand Formation fly-past in the 1932 NAD display led by the HP W.10 with, on its left a Tiger Moth and Fox Moth and on its right, DH.60 Moths, a Desoutter, an Avro 504K and Cierva C.19 Mk IVP. In the top foreground are a Spartan Three-seater and an Airspeed Ferry.

continuing to court Lord Wakefield's patronage regarding oil supplies Cobham took the calculated risk of alienating Shell by persuading National Benzole to become his fuel supplier for the next four years.

This was just one of a myriad of similar organisational details occupying his mind as he attempted to focus his entrepreneurial thinking on the Mandate, which appeared on the display programme in the following abridged form:

"I want at least a million people to append their signatures to a Mandate for British Aviation as an expression of their desire for a more energetic development of flying.

"The Mandate has been compiled by the eleven leading aeronautical organisations, each of which has contributed a clause dealing with an urgent necessity lying within its own particular province. As each Society is the recognised authority in its individual sphere the Mandate, as a whole, is a complete and authoritative declaration on the action recommended for the success of British aviation.

"At the end of the campaign it is my intention to place this Mandate, supported by the million signatures, before the proper authorities and municipalities as an indication of the strength of public opinion in favour of a more progressive policy for civil aviation.

"The entire campaign is being conducted under the auspices of the principal aeronautical organisations and it is my earnest hope that their members will assist in obtaining signatures from the public assembled at my various displays, and by private canvas in their own circles.

I am confident that if every opportunity is taken to place the Mandate and its objectives before the nation the number of signatures will reach the million mark."

In order to give Cobham's National Aviation Day display a touch of class, it was decided uniforms would be worn by the flying personnel. However, these proved to be uncomfortable and too hot when worn throughout the long summer day. They were soon rejected in favour of more practical cool white overalls sporting golden brevets, along with black ones with half-brevets for the ground engineers.

Cobham, fully aware of the paramount importance of publicity, was nevertheless taken aback when he realised how many thousands of greater than life-size posters featuring his image and the caption "Fly with Alan Cobham" had been posted without his knowledge on seemingly every billboard, house wall and

available public surface through the United Kingdom. He later recounted how it all again came to light in the early part of the Second World War. Required by Lord Beaverbrook to undertake a tour of the British Isles searching for Alternative Landing Grounds, he was constantly confronted by the famous poster that had, with varying degrees of success, withstood the onslaught of wind and rain over the intervening years.

As the 1932 tour got underway, Cobham became increasingly worried about the drain on his finances. Rain affected many of the early displays and the £300 that it was necessary to take each day to break even seemed a distant goal. But, at Walthamstow, some two weeks into the tour, things started to look up – not least the paying spectators and the display then began to turn a mild profit. Inevitably, accidents requiring Cobham's attention also began to happen. At Newcastle, pilot Turner Hughes' momentary lack of discipline during the propeller-swinging start up of a DH.9, left Percy Allison and his ground crew companion with several broken bones between them. At Blackpool, just days earlier, on 27th June, Allison had been involved in another unfortunate incident. Having suggested to Captain Fielden that a downwind take-off with a fully-laden Ferry G-ABSJ was, to put it mildly, 'inadvisable', he could only await the inevitable outcome when the pilot decided to go ahead. The damage resulting from the aborted take-off was largely confined to the aircraft itself, although two passengers also sustained minor injuries. Cobham, however, regarded the incident as grounds for Fielden's dismissal, whereupon Hugh Johnson became the new chief pilot.

Despite his impressive reputation and experience, Captain Percy Phillips was also called to account by the fiery Cobham when he landed in the path of another machine. But on this occasion, things were not so easily resolved for, despite the leader's warranted criticism, Phillip's staff, loyal to a man, successfully demanded that Cobham withdraw his comments. Very few however, regardless of seniority, were favoured with an apologetic retraction following a blistering confrontation if Cobham considered that some important administrative detail had been overlooked, or that work was seen to have been improperly carried out.

Certainly it was woe betide any employee referring to the display as the 'flying circus' within Sir Alan's hearing. Although a popular sobriquet adopted by the media and public alike, Cobham considered that the term undermined his serious attempt to instil a 'professional' attitude towards aviation. Nevertheless, it was, as he later reluctantly admitted, a situation that he had to accept with seeming good grace.

'Circus' was, however, a term particularly suited to describe certain comic moments that occurred on the tour. The display

Comper Swift G-ABPY, which suffered a runaway at Liverpool in June 1932.

The National Benzole refuelling truck which doubled as a defuelling facility for W.10 pilot Hugh Johnson!

programme's list of events usually stated that a 'surprise item' would be included, but on one particular occasion the crowd certainly got more than it bargained for. An aircraft, having completed its aerobatic routine landed, but promptly ground looped causing it to demolish a tent and to run into several official cars parked nearby. The crowd then let out a huge cheer at what it thought was a sensational piece of 'surprise' showmanship!

Annoying and careless incidents were sometimes caused by experienced pilots who, it might be said, ought to have known better. On more than one occasion, no doubt due to the pressure of keeping the show on schedule, a pilot attempted to start his engine unaided, only to have the unchocked machine run away before he could enter the cockpit. At Liverpool, on 11th June 1932, a disaster was only narrowly averted when ground engineer Williams managed to seize the wing-tip of an unoccupied Comper Swift G-ABPY, causing it to buzz around like an angry bee, the pilot, as reported at the time 'attempting all the while to enter the mad circle like a child trying to enter a skipping game'. Forced in the end to release his grip, Williams could only close his eyes in despair as a loudspeaker pole brought the Swift's unmanned progress to a sudden halt. Needless to say the paying public enjoyed the impromptu performance immensely, whereas an infuriated Cobham most certainly did not.

The demands placed each day on the joy-riding pilots were indeed formidable. There was no time formally set aside for breaks of any description and Hugh Johnson, who was usually flying the W.10, had to settle for sandwiches and a paper cup of coffee which he drank as and when the opportunity arose. The long queue of passengers rarely allowed him the luxury of leaving the cockpit in search of the nearest 'convenience'. This particular problem was solved however when Johnson persuaded the driver of the National Benzole refuelling truck to let him defuel himself through a gap in the cabin floorboards with not a minute wasted!

It also wasn't long before Sir Alan's personal charisma and that of his flying team began to attract a strong female following. It fell to Cobham's secretary, Frances Cameron, to point out to her invariably pre-occupied boss that the same girls who turned up at successive venues were in fact not aviation enthusiasts as such but were being disported daily in the W.10 'concubine carrier' as a tour 'morale booster' to keep the boys happy!

The passenger-carrying pilots were called upon to fly long hours each day, often well after dark with only lorry headlamps to

Turner Hughes' remarkable aerobatic performances were undertaken in the distinctive yellow and silver Tiger Moth, G-ABUL.

provide a crude aid to landing, but inevitably it was the more spectacular aerobatic pilots who captured the public's undiluted admiration. The star attraction was the tall, debonair Charles Turner Hughes ('Toc H' to his friends) who, flying a specially modified Tiger Moth G-ABUL, completed no less than 2,328 loops, 2,190 rolls, 567 stunts, 522 upward rolls, 40 inverted falling leaves and 5 inverted loops in the year he spent with Cobham. During this same period, 'Toc H' made 3,830 landings and spent some 170 hours inverted. When considering the concentration required to position his aircraft such that the crowd was not blinded by the sun, or required to bend necks beyond a forty-five degree angle, Turner Hughes' Herculean performances deserved and received the highest accolades.

Martin Hearn, a member of the Northern Air Transport touring contingent that formed part of Cobham's travelling troupe in 1932 along with the Cornwall Aviation and Rollason teams, soon came to be recognised as a very special performer whose 'wing-walking' antics were simply not to be missed. Such was his fearless willingness to 'dice with Jesus', as he so prosaically described it, that he had to be forcibly dissuaded by Cobham from carrying out a loop while standing virtually unsupported on the top wing of an Avro 504. Reluctantly accepting that this might be pushing divine indulgence too far, he settled for undertaking the manoeuvre sitting on the Avro's central skid just forward of the main undercarriage!

A polished performance was invariably provided by Charles Herbert Lowe-Wylde, known to all as 'Jimmy', in his BAC VII two-seater glider. Lowe-Wylde was an accomplished aeronautical engineer who, having founded The British Aircraft Company at Maidstone in 1930, (it became BAC Ltd in 1931), justifiably

Assembling Jimmy Lowe-Wylde's BAC VII glider for a 1932 NAD display.

claimed that there was no part of any of his machines that he could not produce with his own hands. He also pioneered the technique of launching gliders by towing them behind motor cars and gave demonstrations of this at several NAD displays in 1932.

The success of his glider designs led to the fitting of a Douglas 600cc motorcycle engine on a steel pylon above the wing of his BAC VII glider. It was while flying the prototype (BGA 194) of this ultra-light machine, now called the Drone, that Lowe-Wylde became ill and suffered a fatal crash on 13th May 1933 at the Maidstone Flying Club at West Malling in Kent. Following his death another influential member of the British gliding movement, Robert Kronfeld continued to produce the Drone.

Another crowd-pulling event was the parachute drop which, though usually performed with pin-point precision by either Captain EW Stewart or Ivor Price, certainly carried with it a high degree of risk. The crowd's sense of anti-climax following a safe descent was usually all too evident but,

as will be seen, the thrill-seeking public did on several occasions witness tragedy at close quarters.

As the 1932 tour drew to a close, an unusual accident occurred on 5th October at Harrogate. Captain ED Crundall, an experienced Cornwall Aviation pilot, came to grief when, despite requests by Cobham to carry out his manoeuvres in Avro 504K G-AAUJ at a higher altitude, he still continued to cut his safety margins too fine. When the inevitable crash came, just out of sight of the crowd, Turner Hughes was instructed to distract the crowd with a show-stopping display of aerobatics. Cobham meanwhile, with Airspeed colleague Neville Norway (later the novelist, Neville Shute) and his wife Dr Frances Norway, raced across to the crash scene but were unable to save the life of one passenger, Mr Arthur Snow. According to *"The Aeroplane"* magazine's report, the other occupant, a Mr George Walker, immediately took another flight to 'steady his nerves'.

Fortunately, the only aviation journalist to witness the event was a good friend of public relations manager, Edwin Rossiter, and the sad affair was largely passed over. The accident was, however, reported to the Air Ministry and at the subsequent inquest, the jury, having been asked to note that there were a number of causes, apart from any incorrect action by the pilot, that might have caused the crash, returned a verdict of 'accidental death'. An added statement however, pointed out that Air Ministry instructions regarding the wearing of seat belts were too vague. Although it was already a compulsory requirement to provide seat belts, it was highly unusual for passengers to wear them and no pressure was exerted upon anyone to do so by operators keen to keep turn-round times down to the minimum.

At the end of the summer, Cobham was tempted by offers of up to £40,000 to sell his organisation, but mindful of Leslie Castlemain's visit to South Africa with the Spartan Air Circus the previous winter, he also considered the prospect of capitalising on his own earlier headline-catching flights to the Union. Within days his mind was made up – the attraction of another African challenge having proved irresistible.

Martin Hearn 'wing-walking' on a Cornwall Aviation Avro 504K at an NAD display at Roborough aerodrome, Plymouth on 20th August 1932. This performance was unsurprisingly banned the following year.

The DH.66 Hercules G-ABMT took the place of the HP W.10 in the South African tour and carried the names "City of Cape Town" on one side of its nose and "Stad van Kaapstad" on the other side.

Chapter Five: Southern Interlude

*'The most positive men are the most credulous,
since they most believe themselves'*

Pope – Thoughts on Various Subjects

Cobham's decision to take a touring display to South Africa placed an enormous and immediate burden on his administrative lieutenants, the main responsibility falling on Edwin Rossiter. Once his duties surrounding advance publicity for the 1932 UK tour had eased, he immersed himself in the myriad details associated with the packaging and crating of aircraft and equipment, ground transportation and of course the travelling and accommodation requirements of the touring party. This task completed, he then took on the running of the London office, putting in place the forward arrangements for an even more ambitious traverse of the British Isles by Cobham's Display in the following season.

Sir Alan meanwhile had managed to secure Imperial Airways' promise to detach its DH.66 Hercules airliner, G-ABMT, *"City of Cape Town"* airliner from its normal duties for the tour. He also secured a commitment from Sir John Siddeley to loan an Armstrong Whitworth AW.16, G-ABKF (soon dubbed the *"Red Devil"*) and several saloon cars to complement his new sponsor Leyland Motor's provision of heavier transport. Both road vehicle suppliers agreed to deliver their charges in a creamy white finish that Cobham thought most likely to retain a consistently smart appearance in Africa's demanding conditions.

Ahead of the main party, Cobham sent Leslie Castlemain and chief inspector Dick Goodban armed with maps and a rough outline of what was expected of them once they reached the southern shores. After disembarking at Cape Town, they carried out a general reconnaissance, travelling clockwise around the Union, and managed to secure the co-operation and willingness of no less than seventy-seven town and dorp (village) authorities to stage a display. The broad and flat areas usually available meant that the

problems previously encountered in England regarding the preparation of landing grounds were no longer a serious cause for concern.

It soon became apparent, however, that the inadequate cross-country road system with its unending succession of deep gullies, and the tendency for flash floods to spring seemingly from nowhere would bring its own special difficulties. Additionally, great reliance was placed on local prison governors to provide working parties to clear the many rocks and stones which invariably littered the sites earmarked for the shows.

The rules and regulations then existing in South Africa did not normally allow companies to transport their own goods from one location to another, such movements having to be undertaken by the monopolistic state-owned railway company. However, following Castlemain's determined efforts, Parliament passed a special Act which afforded Cobham a unique dispensation, allowing his convoy to proceed under its own steam. With this formidable hurdle out of the way and the tour's administrative framework in place, a larger party comprising Dallas Eskell, Sammy Sowerbutts, Hugh Johnson and Lady Cobham accompanied by her two young sons Geoffrey and Michael set sail on the *"Warwick Castle"* in order to consolidate the final arrangements.

Shortly afterwards, Sir Alan and his pilot companions also departed for the Union from London's Victoria Station, but not before bidding farewell to his mother and father (now aged seventy-five and in failing health). As efficient as ever, Rossiter placed in Cobham's care, a briefcase containing every conceivable item of

An aerial view of Cobham's display at Durban 28-29th January 1933 with its rudimentary enclosure and many non-paying spectators visible outside!

Flash flooding brought many difficulties for the South African tour's ground transport.

In addition to swollen rivers, the rough terrain and poor road conditions played havoc with ground transportation on the South African tour.

The Armstrong-Whitworth AW.16 G-ABKF was flown by Turner Hughes in South Africa but had to return home early with engine problems.

essential information regarding men and machines accompanied by the good wishes of the staff members remaining at home.

Immediately on his arrival in South Africa, Cobham set about producing a Mandate based on the one which formed the *raison d'etre* for his English 'air-mindedness' venture. Printed in both English and Afrikaans it again underlined the ability of the aeroplane to improve communications over long distances and the need for improved airport facilities to be provided on a grand scale – a message enthusiastically endorsed by South Africa's own flying hero Sir Pierre van Ryneveld. Not surprisingly, it also stressed that in African aviation's bright future the tour's main sponsors, Imperial Airways and Armstrong Siddeley hoped to play major roles.

And so, a mere six weeks after the NAD's 1932 programme had ended, Cobham's first South African air show opened on 1st December at Cape Town. It featured the DH.66, three Avro Tutors, G-AARZ, G-ABZP and G-ABZR, a Cierva C.19 Mk IV Autogiro G-ABGB and the AW.16. Also attached to the fleet was an Avro Avian belonging to the Vacuum Oil Company which was contracted to supply all the petrol and oil for Cobham's aircraft.

No doubt prompted by the free flights sponsored by the Cape Times, the public response was excellent. However, when Cobham realised that no aircraft had ever flown out of Cape Town after dark, he seized upon the additional novelty of providing views of the city and Table Mountain from the air at night. This task naturally fell to Hugh Johnson who, after each evening's meal and still wearing his dinner jacket, proceeded to undertake several trips

from the airport's circular grass field which, to his consternation, boasted merely a single searchlight as landing aid equipment.

Although the flying events proved highly successful, the parachutist Ivor Price encountered problems when, still jumping from his accustomed height of 4,000ft, he found that the thinner air at higher venues on the High Veldt accelerated his descent. Despite being knocked unconscious on more than one occasion, he nevertheless survived the tour with body and enthusiasm remarkably intact, to appear on Cobham's UK excursion the following summer.

Not unexpectedly the question of racial segregation raised its head whenever large crowds gathered. Refusing to share joy-riding flights with blacks or coloureds, 'special' performances had to be provided for white customers only, in order to satisfy the local authorities. Cobham later observed that, despite the great disparity in wealth, it was invariably the white community which remained outside the primitive screening for a free show while the less privileged social classes stoically stood in line to pay the entrance fee.

Day-to-day administration of the tour proved far from easy and tensions continued to simmer between Cobham and tour manager Dallas Eskell. These feelings soon erupted when it transpired that scant regard had been given to the sun's location in the Southern Hemisphere and, as a result, at the first show the crowd would have been staring into the blazing sun had it not been for Cobham's last minute changes to the siting of the public enclosure. Eskell's subsequent reprimand, though certainly delivered in no uncertain terms, bore no comparison to the verbal barrage Cobham launched at Turner Hughes when, following an engine seizure, it was discovered that he had neglected to top up the *"Red Devil's"* oil tank. With the engine bearings ruined and no replacements available, both 'Toc H' and the aircraft were forced to return home.

The Cierva autogiro also threw up many difficulties for its pilot, AC Rawson. Unless the machine was allowed to operate from an absolutely flat surface, a condition that could rarely be guaranteed on this or even the UK tours, the rotating blade assembly had a disconcerting tendency to become unbalanced. This caused the aircraft to rock from side to side in a circular movement that became referred to as the 'Dance of Death'. Both pilot and passenger could then only sit, crouched as low as possible in their cockpits, to await the eventual slowing down of the scimitar-like blades whirling just four feet above their heads. Although clearly perilous, the 'Dance of Death' resulted in total structural disintegration on only two occasions during the four years this type

After suffering the loss of the Cierva C.19 autogiro G-ABGB, Cobham was fortunate in being able to acquire a local replacement.

Illustrative of the effects of the 'Dance of Death' from which early autogiros were prone to suffer; this one was the result of a similar event at Apse Manor on the Isle of Wight.

of aircraft was featured in the display. The first such incident did, however, take place on 13th February in Cape Town and in view of the autogiro's undoubted crowd appeal it was particularly fortunate that a local owner was persuaded to sell his machine, G-ABFZ, as a replacement, and in the light of his unnerving experience, that Rawson was prepared to fly it!

Shortly after this escapade, one of the Avro Tutors required repair when its pilot, Ogden, misjudged his take-off and deposited the aircraft at the bottom of a ravine – fortunately without damage to himself.

Hugh Johnson later recounted another event that also nearly had disastrous consequences. Flying the DH.66 over a valley bounded by four large hills he encountered a freak weather condition which caused his machine to suddenly plummet several thousand feet within a few seconds. Passenger seats were torn from the floor attachments throwing the occupants around and causing them to hit and puncture the fabric inner lining of the fuselage roof. Johnson himself suffered a nasty gash over his eye when the Very light pistol flew out of its holster whilst he was struggling to maintain control. An equally strong upward movement then brought all the passengers hard down onto the floor and, the pilot's injury apart, it was miraculous that no serious harm was caused. Johnson however, was a little miffed when, back on the ground, Cobham casually enquired "You are going to carry on aren't you old man?" as he dutifully applied perchloride of mercury antiseptic to the still bleeding wound!

Johnson's flying skills were further called upon at Robertson, a small township where he found he was required to fly the Hercules airliner from a golf course that featured two fairways running at right angles to each other. To do this with such a large aircraft possessing no brakes proved quite a feat, requiring a full power charge down the first fairway – before throttling hard back on one engine and applying full rudder to negotiate the corner into the other. According to Johnson, little did the passengers know the risks they were paying good money for! He might, however, have considered himself an old hand at dealing with such challenging situations for, in England on another occasion, he had been required to depart round a similar sharp bend with a man at the blind corner armed with red and green flags to signal 'stop' or 'go' as appropriate.

The displays in South Africa ended on a sad, indeed bizarre, note. In England, the flying personnel always attracted a large female following wherever they went and the local girls here were no different. At Pietermaritzburg, Leslie Castlemain had arranged for several socialite ladies to 'entertain the boys' and it was faced with this temptation that Tutor pilot, Harold Lawson, although romantically involved with a girl at home, became infatuated with another. Unfortunately the show's announcer, Ross, proved a keen rival for her affections. The complications of this situation are thought to have brought about the distraught Lawson's declared intention 'to end the tour with a bang'. This prophesy he tragically fulfilled when, on 17th February he plunged his machine (G-ABZR) into the circle marking the centre of Cape Town's Wynberg airport, taking Ross's life in addition to his own.

In February 1933, Sir Alan received news of his father's death and, with affairs at home now demanding his attention, he left the running of the tour to his wife and Dallas Eskell. But a serious situation arose soon after his departure when Eskell developed a mysterious migraine condition that, worsening daily, entirely incapacitated him. Constant medical supervision was necessary and arrangements were made for his repatriation home by sea with Castlemain accompanying him. This proved to be an arduous experience for Castlemain who, on one occasion was required to exert extreme physical force to prevent his charge, by now determined to end it all, from jumping overboard. Fortunately, by the time the ship reached Madeira, Eskell's health had appreciably improved and he went on to make a robust recovery in time for the 1933 touring season.

The three-month campaign which had stretched the organisational capacity of Cobham's team to the limit was not a financial success. Although 13,593 passengers had produced £8,006 in flying receipts, the net result was a loss of £6,000. Nevertheless, Cobham correctly foresaw that it did condition the attitudes of many young men and encouraged them to fly in the war that within a few short years, would boil over to engulf the world.

Harold Lawson is presumed to have committed suicide at Cape Town during the NAD tour.

With the Handley Page Clive landing, shortly followed by an Airspeed Ferry, this picture shows how easily accidents could occur, such as at Leeds in June 1933, with a public unfamiliar with aerodrome control procedures.

Chapter Six: That's Entertainment

*'Now Cobham said he wanted a lad with lots of pluck
(You needed it in those days and also bags of luck),
So flying those fantastic kites, rigged up with old shoe laces,
Our Cecil made his name as one of the Cobham Aces.'*

'Capricorn' 1945
(Supplied by CWH Bebb)

Upon his return to England, Cobham discovered that Henry Barker, his erstwhile National Aviation Day Limited secretary and Jimmy McEwan King of Galbraith King and Company Limited, who previously had undertaken all the printing requirements for his shows, had set up their own business under the banner of the British Hospitals Air Pageant. The organisational expertise they had gained in 1932 was thus now transferred into a competitive camp whose business aims Cobham soon came to regard as downright fraudulent.

In considering ways to counter the threat that the BHAP posed to his own forthcoming display attendances, he decided to promote two NAD tours that would operate concurrently at both ends of the country. The finding and approving of over 350 landing grounds for this combined operation proved to be a mighty task, but by the middle of April 1933, all the arrangements were miraculously in order with new pilots and ground staff engaged and additional aircraft purchased, overhauled and freshly certificated.

Both displays progressed gradually northwards, with Eskell's No.2 tour concentrating mainly on locations on the western side of the country, and Cobham's No.1 tour on those on the east. All went well until Cobham arrived for a two-day stay at Middleton Park in Leeds on 11-12th June. It was here that, despite several warnings, two young boys, Fred Smith aged twelve and Leslie Taylor aged eight, rode their bicycles under the Handley Page Clive airliner, G-ABYX, as Hugh Johnson made his landing approach. Both boys were killed outright, but the subsequent court of inquiry cleared Johnson of any blame. Cobham remarked in later years how his legendary good fortune again came to his aid on this occasion, for on the boys' headstone in Leeds' Beckett Road Cemetery, it ascribed the accident to an airliner belonging to a "Sir Richard Cobham".

Despite the shadow cast by this unfortunate incident, the No.1 tour, with Cobham's old colleague Fred Holmes not only supplying aircraft, but acting also as chief inspector, soon proceeded to Ireland where its arrival attracted a mixed reception.

This was not the first organised tour to promote the adventures of joy-riding in Ireland. Only the year before, in 1932, Waterford-based Irish Air Lines Limited had visited thirty towns in the Emerald Isle, employing a Blackburn Bluebird IV, G-ABJA (later EI-AAO) and two Avro 504Ks, EI-AAM (ex G-AAYH) and

Avro Tutor G-ABZP rejoined the UK Cobham touring fleet following its return from South Africa and was flown in the 1935 displays by Geoffrey Tyson.

EI-AAN (ex G-ABHP) purchased from Northern Air Transport Ltd. This venture had ended after the Bluebird and one of the Avros were written off in accidents and the company went into liquidation.

Cobham's National Aviation Day touring troupe had, however, been expressly invited by Hugh Cahill, the founder of Iona National Airways, and was the first to venture from the mainland. Prior to its departure, Cobham had delivered a firm warning to all personnel about the dangers of introducing politics or religion into any conversation, particularly in the pubs and similar establishments. It was also necessary, he had added, for everyone to be careful when choosing temporary accommodation in Dublin, pending the arrival of the tour's tents and domestic equipment.

Whilst pondering these pronouncements, Harry Morris, then an inexperienced apprentice accompanying the Holmes' Air Travel detachment of Avro 504Ks, G-ABVY, G-ACCX and G-ABVH overseas for the first time, settled back in the cockpit to await take-off, barely visible under a pile of loose items. But as he was to tell later, his puzzlement at what seemed to be the last minute addition of an inflated bicycle tyre inner tube was eased when Holmes shouted above the engine noise for him to hold on to his 'life belt' as it might come in handy should mishap occur over the Irish Sea. Then, with paraphernalia stored in every nook and cranny, all the display aircraft slowly climbed away and turned westward to follow Bill Kingwill's Avro, which was the only machine equipped with a compass!

The first port-of-call was Dublin's Phoenix Park. Here an established gang of vociferous women, who, by tradition were accorded the sole rights to sell programmes at any public event, immediately made it clear to the Ketchell brothers that their presence was not required. What was required however, was for Ketchells' stock of programmes to be sold at a nominal price to the ladies who threatened rapid emasculation unless they complied!

Once this small technicality had been resolved, the display got underway with Martin Hearn soon beguiling the crowd with his 'wing-walking' antics. Following a well-established routine, Wing Commander Tresillian Bulmore, the new announcer, then called over the loudspeaker for any volunteers willing to match Hearn's performance. After a period in which no-one stepped forward, a strange looking apparition in a comic suit declared his intention to 'have a go!' With much cheering and clapping, he was given some exaggerated instructions before the aeroplane took off. It was to reappear soon after, now some fifty feet above the crowd with the brave volunteer seemingly making a complete hash of climbing out onto the wing. After a couple of extra low passes over the public enclosure, the next, with the 'wing-walker's' legs flapping frantically below the fuselage, brought gasps from everyone when the figure was seen to fall to the ground. Whilst this was, in fact, a rubber dummy, it appeared so realistic that a priest ran out to deliver the Last Rites and a highly pregnant lady watching the performance from the back seat of a car, promptly went into premature labour. Following this climactic consequence, Cobham decided that this particular event was probably in bad taste and forbade it for the rest of the visit. It is not recorded whether it was re-introduced once the shows were resumed in England!

It was during 1933 that, following a series of other potentially dangerous incidents, unrestricted 'wing-walking' was officially banned. It is perhaps surprising that the authorities had not instigated this earlier, for in 1929, Air Ministry Notice to Airmen No.41 had stated:

> "The attention of all concerned is directed to the structural damage which may be caused through demonstrations of 'wing-walking' being carried out during flight on light aeroplanes such as the de Havilland Moth, Avro Avian, Blackburn Bluebird, Simmonds Spartan, etc.
>
> If the practice of 'wing-walking' on the type of aircraft referred to above is pursued, it will be necessary for the Air Ministry to prohibit it under Article 9 (4) of the Air Navigation (Consolidation) Order, 1923."

Hugh Johnson recalled that the Guinness tent played a crucial part in encouraging the Irish crowds, which always contained a high proportion of priests, to fly. As the evenings wore on, many would- be aviators, by now relaxed and in good voice, had to be forcibly persuaded that sitting on the undercarriage for a flight was not a good idea. Again, however, as in Cape Town, the temptation to swell the 'circus' coffers with revenue from night flying proved

Air Travel's distinctive Avro 504Ks formed part of Cobham's contingent to Ireland in 1933.

One of Geoffrey Tyson's favourite mounts was this DH.83 Fox Moth.

irresistible. On Johnson's ears fell Cobham's pleas to take on "just another load or two, old man". This always meant that a substantial number of customers were still waiting their turn. Flying often continued until well after midnight, by which time Johnson's ability to concentrate on landing aided only by the lights of the National Benzole fuel truck, was understandably wavering.

Certain sections of Dublin's population were, however, less than enthusiastic regarding any visitors from the mainland. Signs permanently daubed on walls demanding the rejection of British goods were soon accompanied by others calling for all to "Boycott Cobham's Air Display". It was even proclaimed in the press that the air circus was merely a pretext for surveying the area prior to a future aerial attack.

Unsurprisingly, many members of the touring team suddenly brought face to face with such extreme anti-British feeling, wondered whether Sir Alan had not made a serious and provocative error of judgement. Though paradoxically passenger-carrying continued to be brisk, a general feeling of unease seemed to accompany the tour.

More trouble soon presented itself. On 7th July 1933, at Ballycummin, Limerick, Geoffrey Tyson, recently elevated to chief aerobatic pilot following the departure of Charles Turner Hughes, lost the undercarriage on his Fox Moth, G-ABWF when it was struck by a DH.60G Moth, EI-AAI flown by Captain William R Elliott, chief instructor of the local Irish Aero Club. Elliott's unauthorised presence in the display area cost him his life and that of his passenger, William Ower. Tyson landed safely, but was not spared blame for the accident by a section of the crowd.

The tour personnel were advised by the Garda to keep a low profile and not to venture away from the field that night. This turned out to be wise counsel for, after dark, a large number of incensed town folk bearing lighted torches and shouting obscenities were only prevented from setting fire to the whole of Cobham's outfit by the local police who performed with great courage and determination. Although the accident received full coverage in the leading press, political considerations resulted in no mention being made of the subsequent crowd disturbance.

For Cobham, the problems encountered in Ireland appear not to have been too daunting, for less than two months later, he returned for a repeat performance. This time all went well until the day before the display was due to return to England. On 1st October 1933, Colpe Farm, Drogheda was the site of a fatal crash involving a Spartan three-seater, G-ABZH piloted by Captain K Rose. This was one of two machines (the other being G-ABYH) that Aerofilms Limited used to provide photographic coverage of the Cobham tours in 1933. On this occasion, the Spartan was also giving joy-rides and although Captain Rose was killed, his passengers, Mr and Mrs Roe survived. Understandably, this sad experience did not encourage Aerofilms to support the display tours so robustly in subsequent years.

Twelve aircraft accompanied each of the No.1 and No.2 tours. In addition to the usual mix of Avro 504s, DH Moths, Tiger Moths and Fox Moths, included in each complement of machines was a Blackburn Lincock used for demonstrating 'radio-controlled' aerobatics. A Mk.II version, G-AALH, was flown by Geoffrey Tyson and a Mk III, G-ABFK was put through its paces by Cecil Bebb. In view of the public's demand for flights in the HP W.10, Cobham had also purchased another ex-Imperial Airways airliner, the twenty-two seat Handley Page Clive, G-ABYX which, though firstly named *"Youth of Australia"*, later became more familiarly known as the *"Astra"* and was piloted by Hugh Johnson, who, in addition to a retainer fee, received a commission based on the number of passengers flown.

Johnson developed the trick of advancing the throttles of the open exhaust engines when taxying close to the announcer's microphone. Making such a thorough and exceptionally noisy nuisance of himself in this manner ensured that the announcer concentrated his efforts on getting Johnson's machine full of passengers before the other joy-riding pilots received their quota.

Although the airliners were used almost exclusively for the more sedate passenger flights, an alarming incident occurred when, over Uxbridge, Martin Hearn was allowed to try out a trapeze act suspended some sixty feet below the *"Astra's"* fuselage. All had gone well until Cobham and another crew member tried to winch him back on board. It was then that the steel cable attached around Hearn's waist slipped off its pulley and became jammed in the winch cogs. Alarmed passengers, keen to assist, merely exacerbated the problem when their combined weight acting toward the rear of the machine caused the aircraft's nose to rise. Cobham had to push them back in order for Johnson to restore control. With seemingly little chance of getting Hearn to safety and the aircraft now dangerously low on fuel, the situation looked bleak, but Cobham tied a heavy spanner to a rope fortuitously lying nearby in the cabin, dangled it down to Hearn and indicated that he should pass it round his chest and to hold on tight. It then proved just possible to pull him back on board while Johnson lined up for his landing approach. It had been a close call and the 'man on the flying trapeze' act was never tried again.

It was Johnson who later unwittingly proved to be the 'villain of the piece' when 'jumper' Harry Ward came within an inch of losing his life. It was a standard feature for the *"Astra"* passengers to be given a close up view of a descending parachutist but at Weymouth, on 23rd August 1933, the risk involved became all too clear. As the *"Astra"* circled ever closer to Ward, the turbulence produced by the two 525hp Bristol Jupiter IX engines turned Ward's Russell Lobe parachute inside out and into what he later described in his autobiography *"The Yorkshire Birdman"* as 'an untidy bundle of laundry'. His precipitate contact with the ground resulted in a broken ankle, dislocated hip and severe concussion.

Ward was, as he described it, 'in a right mess' both physically and financially for it was impossible in those days to obtain any kind of insurance for such high-risk occupations. But, irrepressible as ever, upon being told by Cobham that he should have made proper provision from his retaining fee of £16 per week and £2 per jump allowance, Ward directed the bill for his month long stay in hospital to NAD's London headquarters. He heard no more about it!

When told by medical staff "You'll never jump again" as he hobbled out of Weymouth's General Hospital, Ward replied "we'll bloody well see about that" – a response typical of the feisty breed of parachute performers that also included a remarkable one-armed exponent, Robert Wyndham. Ward's subsequent career with other flying circuses and the part he eventually played during the Second World War in founding the RAF's Parachute Training School, are ample testament to his fortitude and determination.

On tour it was customary for Cobham and Eskell to have a telephone discussion at the end of each day to assess, amongst other things, their respective takings. To prevent anyone else overhearing the actual amounts a code was used in which a farthing represented £25. Hence, for example, 'sixpence' (2½p) which consisted of twenty-four farthings equated to £600. Usually the amounts involved were of this order, but on one particularly memorable occasion after a two-day period of exceptional weather with the No.2 tour ideally situated between three large Yorkshire

Aerofilms Ltd had a monopoly for photographic coverage of the NAD tours but their interest waned after the fatal accident to one of their Spartan Three-seaters, which they used for air-to-air photos.

The purple-painted Blackburn Lincock G-AALH was guaranteed to give a thrill to the crowds when flown by aerobatic pilot Geoffrey Tyson.

towns, Barnsley, Sheffield and Huddersfield, the receipts amounted to 'two shillings and five pence three farthings', i.e £3,000 – a huge increase on the average 'take'. Barely able to contain himself whilst informing the boss of his good news, Eskell's expectation of fulsome praise was considerably dented when Cobham, ever the pragmatist, merely replied "By God we need it!"

In June 1933, the No.1 Tour was putting on a display at Cambridge when, from out of a visiting DH Moth stepped a slim and attractive girl. Very few women operated on the professional flying circuit at that time and Cobham quickly recognised the newcomer to be Jean Batten, who was soon to earn international fame as New Zealand's long-distance record-breaking heroine. Though keen to accede to her request to join the touring team, Cobham, after discussions with Tyson and Johnson, sadly had to refuse. Both his senior pilots, though impressed with her general flying abilities, had expressed concern that her limited experience

of flying out of small restricted areas might place a burden on others at critical moments. Meeting again many years later, she playfully chided Sir Alan by telling him he was the only man who had ever turned her down!

Sir Alan's intention to wage war on the 'free loaders' who evaded the official points of entry to the display amounted almost to an obsession. Always on the look out for new ways to disperse such gatherings, he hit upon the idea of sending out ground staff on motor cycles to make their presence as uncomfortable as possible.

Terry Brien recalled the occasion when, on 24th April 1933 at Alton, Hampshire, he was concentrating on such a task and to his amazement, as he approached the ranks of unwelcome guests, he saw that instead of moving sharply out of his way many were throwing themselves flat on the ground. A quick glance over his shoulder revealed the undercarriage skid and whirling propeller of 'Jock'

Mackay's red Avro 504K, G-EBIS almost upon him. The noise of his motorcycle exhaust had blanketed out the approach of Mackay's aircraft flying at head height on the same mission. Although lowering himself rapidly to one side prevented his immediate decapitation, one of the 504's wheels removed the motorcycle's headlight and handbrake lever. It was only when Sir Alan accompanied him back to the scene of the incident that the boss, until then convinced that the damage was the result of simple carelessness, realised that a bizarre accident had been so narrowly averted.

Despite the competition provided by the British Hospitals Air Pageant, the 1933 National Aviation Day tour proved very successful. Cobham's separate teams visited a total of 306 towns in the British Isles and close on a quarter of the 800,000 people that paid for admission had gone on to enjoy a pleasure flight. Twenty-two pilots and some 100 ground staff had been employed throughout the season, which, though bringing a number of close calls, had not seen a flying passenger harmed in any way. To celebrate this, Cobham arranged a *fin de saison* display on 7th

A scene that well illustrates the nomadic existence of the Cobham displays.

The tented accommodation which was used on the Cobham tours was managed by specialist attendants. Note the leaded windows to the Pullman-Read portable restaurant!

Above: Hugh Johnson exercising his skills to ensure a regular supply of passengers for the HP Clive.

October in which machines of both No.1 and No.2 'squadrons' met up over the London head office before proceeding to the winter base at Ford.

As 1933 drew to a close, Cobham decided that the following year would see a reversion to a single tour, but one that would encompass extra new towns. To this end he retained his three leading pilots, Geoffrey Tyson, Cecil Bebb and JD Parkinson to carry out a thorough aerial survey for more appropriate landing sites. This obvious, cost-effective approach cut down considerably the areas that had to be later concentrated upon by the teams negotiating with landowners during the winter months.

Although Cobham's touring campaigns made big demands on his time and energy, he also devoted considerable attention to a projected non-stop flight to India, flying an Airspeed Courier. Accordingly, during the winter months of 1933/34, preparatory work was undertaken with aircraft from the NAD fleet modified to perform the task of air-to-air refuelling.

As the summer of 1934 approached, Cobham and his team of stalwarts, now re-formed as a new company, National Aviation Displays Ltd, set off round the country yet again, this time in vehicles provided by Ford. Though involving fewer aircraft, this display featured 'The Three Aces', a formation aerobatic team of three Avro 640 Cadets led by Cecil Bebb. The aircraft were distinctively painted, red (G-ACLU), white (G-ACPB) and blue (G-ACOZ) and demonstrated not only synchronised manoeuvres but pylon racing with thrill-seeking passengers well strapped in. Throughout the season, Geoffrey Tyson continued to carry out precision flying at its best, including his *coup de theatre* – the picking up of a handkerchief from the ground with a hook mounted on the wing tip of his Tiger Moth, G-ACEZ. This he performed twice a day, seven days a week.

Tyson also perfected the stunt of diving down to pass between two poles barely a wingspan apart, between which was stretched a rope with flags attached. Immediately after 'threading the needle' he performed a vertical loop, before passing under the rope again –

Parachutist Harry Ward, pictured here with the 1933 NAD tour.

Robert Wyndham lost an arm in a flying accident but continued to work as a display parachutist.

Another view of the Grand Formation fly-past with the inevitable HP Clive in the lead, followed by a gaggle comprising Avro 504K, Tiger Moth, Fox Moth, Avro 504K, Spartan Mk II and Cierva C.19 Mk IVP.

A graphic illustration of the incident where Cobham ground assistant Terry Brien just avoided decapitation by Jock Mackay's Avro 504K.

The 'Three Aces' team of Avro Cadets was led by Cecil Bebb in the blue-painted G-ACOZ.

usually with barely inches to spare above the ground. Later in the season, Cobham witnessed Tyson's undercarriage actually compress slightly as his wheels brushed the grass and then insisted that the act was not to be repeated until a more powerful machine was available which would allow a tighter manoeuvre to be made. Even *"Flight"* magazine's editor recorded that 'in his efforts to provide a good show, Tyson does incur some serious risks'. Fairly mild comment perhaps in light of the fright he gave Cobham!

Although some flexibility was retained regarding the price of flights, ten shillings (50p) was the normal cost for a seat in any of the aircraft participating in the opening Grand Formation Flight, or any of the other set programme events in which the public were invited to take part. The exceptions to this policy were the £1 charged for a 'full aerobatic flight', in the Mongoose-engined Avro 504N, G-ACOK by Captain Phillips, who had now rejoined Cobham's organisation, and the five shillings (25p) for short four-minute joy-rides (Phillips reckoned on twelve such flights per hour at peak periods!).

The Spartan Three-seater, G-ACAF, which had performed so admirably in Phillips hands on the BHAP tour, was now flown by Captain EW Jordan. The final pricing decisions were generally influenced by the size of the attendances, which in 1934, with much reduced press coverage of the shows, were now in noticeable decline. There was also a distinct fall off in the attention devoted by local authorities, most evident when, in August, not one member of the Falmouth Borough Council attended the display and its arrival hardly attracted a mention in the local newspaper. Interest did develop however when Hugh Johnson, with Percy Phillips' young son on board, skidded the *"Astra"* into a hedge upon landing at nearby Boskenso.

On occasions, poor weather conditions disrupted the displays' carefully planned arrangements. On 24th May 1934, the Cobham formation was flying toward Scarborough when a sea mist rolled in almost entirely obscuring the ground below. This caused a succession of urgent landings on the fortunately still-visible beach.

However, when a clearing patch of sky later allowed the cavalcade to take-off and land as originally intended on the local race course, the HP W.10 struck a ground support lorry and continued over a ditch, losing its undercarriage and one of its lower wings in the process. A repair team was urgently summoned from Imperial Airways, but when Sir Alan later flew over to inspect the restored machine, his patience and composure were sorely tried when his own aircraft's propeller was smashed on landing!

But while the outcome of these incidents had relatively little effect on the tour, a mishap took place on 30th June that had far more serious consequences. It was at Farnborough on that day that 'Jock' Mackay flying his Avro 504N, G-ACRS performed his 'drunken' flying exhibition once too often. This routine required him, having dressed up either as a yokel or mad professor, to answer the announcer's call for flying lesson volunteers to step forward. He would then enter the cockpit only to fall out on the opposite side. Having laboriously struggled back on board and with a despairing instructor on the ground gesticulating wildly in the slipstream, shouting "Stop that man", he would then take-off in a gyrating series of leaps and bounds. Once airborne he would jettison the 'joy-stick' – a dummy provided for the purpose – before treating the crowd to a display of 'stickless' flying. His *pièce de résistance* was then to climb out of the cockpit onto the Avro's lower wing, the proper control column being held in place by a length of bungee elastic cord. But on this occasion his luck ran out and in attempting to re-enter the cockpit, his leg became entangled with the unusual control arrangement and his low-flying machine hit the ground and disintegrated. Mackay died soon afterwards, but found the breath to offer an explanation for his crash and to put the blame entirely upon himself.

On the brighter side, only a few weeks later, on 25th July 1934, Geoffrey Tyson commemorated the twenty-fifth anniversary of Bleriot's crossing of the English Channel, by flying across the same strip of water in Tiger Moth G-ACEZ but inverted. Having rehearsed the operation during the two preceding days, he had found it difficult to maintain a proper orientation and thus asked

A view of the "Three Aces" with 'Easy' Easdown as wingman piloting the red Avro Cadet, G-ACLU.

The wreckage of Jock Mackay's Avro 504N G-ACRS following his fatal crash near Farnborough on 30th June 1934.

Sir Alan to fly alongside, shepherding him in the right direction during the twenty minute crossing. Upon landing, everyone was aghast at Tyson's appearance for the whites of his eyes had turned bright scarlet. Fortunately they returned to their normal colour within a day or two with no permanent ill effects.

On another occasion, it was a generous mixture of good luck and a cool head that saved Tyson whilst again flying inverted. At an altitude of 600 feet, the pin securing his safety harness worked loose and no longer under control, the Tiger Moth entered a steep dive. By gripping the instrument panel with one hand and the control stick with the other, Tyson amazingly just managed to recover a normal attitude at tree top level.

As the summer progressed, Cobham's thoughts were almost entirely now centred on his forthcoming attempt to fly non-stop to India. To do this required a considerable degree of planning and the conversion of both the display's HP W.10s into tankers for air-to-air refuelling support. One, G-EBMR was despatched to Malta, whilst G-EBMM, now renamed *"Youth of New Zealand"* was earmarked to refuel Cobham's Airspeed Courier soon after it departed from Portsmouth.

However, tragedy attended this venture, for in addition to Cobham having to abort his flight in Malta, G-EBMM suffered a structural failure when returning to join the NAD tour. The crash, on 22nd September 1934 near Aylesbury, killed all four NAD personnel on board, including pilot Charles Bremridge.

Cobham always maintained that the very success of his touring shows sowed the seeds of their own demise. The familiarity now adopted by the public towards his and other touring organisations convinced him that 1935 should see his last 'road show' in the United Kingdom. His thoughts however, still lingered on another venture overseas – this time to India. Far too busy to assess the subcontinent himself, he despatched Lady Cobham along with his ubiquitous pathfinder Leslie Castlemain to sound out the possibilities of a successful tour.

Their final report was, however, discouraging as it was discovered that CD Barnard's earlier visit had created a lot of bad feeling within the administrative community. Poor crowd control appeared to have caused much friction and although Barnard had flown many passengers, it had proved impossible to take any

significant gate money. Vast amounts of litter and debris had also resulted at each flying ground and it was clear that the Indian authorities were not about to encourage a repeat performance.

Cobham therefore abandoned his thoughts of India and, although now already heavily engaged in director's duties with several other concerns, called upon his colleagues to summon their efforts once more for a final British tour that from July 1935 onwards would re-form into two separate halves.

As the year blossomed, a good many of Cobham's old hands left for steadier employment in the aircraft manufacturing industry which was beginning to experience a resurgence. Most of the key flying staff, however, remained to be joined by others that included Flt Lt Leonard Carruthers, an ex-graduate from the RAF's Central Flying School. Called upon to operate in complete contrast to his usual Service type of flying, he soon proved his worth by completing fifty-two flights on his first day and forty-seven the next. This led to eighty-two at Speke soon afterwards, where daily joy-riding continued well past dusk with the aid of car headlights.

1935 was not however, a trouble-free year for the travelling troupe. The tour, which began on 12th April at Fareham, got off to a promising start, working its way uneventfully through the Midlands before departing for Ireland. It then commenced a three-week round of performances billed as the Irish Aero Club – Cobham Tour and included a memorable display at Dublin's Phoenix Park, to commemorate Irish Aviation Day on 12th May under the patronage of Sean F Lemass, the Minister for Industry and Commerce. Perhaps a little surprisingly, Joan Meakin's graceful demonstrations in her Wolf glider, now christened 'Irish Independent' were judged by the media to be the most spectacular event at each location. Certainly a novel feature was her voice being relayed to the crowd as she described her aerobatic manoeuvres by radio-telephone.

Just a few days after returning to the mainland, parachutist Ivor Price, twenty-seven years old and married just two weeks previously, fell to his death at Woodford on 30th May. A veteran of over 800 descents, Price, saving hard to buy a riding stable, always paid meticulous personal attention when preparing his equipment. But on this occasion, an uncharacteristic lapse of attention proved his fatal undoing. Price's normal practice involved laying out his Russell Lobe parachute on the grass well away from the flying

The Wolf glider flown by Joan Meakin in 1935.

The Grunau Baby glider, in which George Collins paid the ultimate price when it was over-stressed and broke up over Ramsey.

area, before collecting all the shroud lines and tying them with a handkerchief close to the canopy, thereby preventing them getting out of place. Possibly distracted by a large number of school children watching him work, Price, on this occasion, neglected to remove the handkerchief, thus making it impossible for the canopy to deploy.

Before the event, Bulmore announced that Price and the display's lady parachutist, Naomi Heron-Maxwell would undertake synchronised jumps from two Avro Cadets. On reaching 2,000 feet both leapt from their aircraft. It was Price's usual procedure to delay opening his 'chute' until he was some 700 feet above the ground, so no concern was registered until the moment when it became evident that he was in serious trouble. With little height to spare, Price's attempt to claw his way up the lines to remove the handkerchief was doomed to failure and he was killed instantly upon hitting the ground. With everyone stunned into silence, the show closed to the sombre accompaniment of the National Anthem.

Despite the depression that descended on all concerned, the underlying resilience necessary to undertake such touring ventures now came to the rescue. The following day, Miss Heron-Maxwell showed great courage when before a large crowd at Retford, she insisted on 'business as usual'. Frederick Marsland, a New Zealander who replaced Price was not a regular 'jumper' but a jack-of-all-trades willing to have a go at anything. Sadly this proved his undoing at a Cobham display alongside the Kingston by-pass on 4 September. Using a Russell Lobe parachute regarded by many circuit 'jumpers' as unsuitable for 'pull-off' descents from the wing, he also fell, enshrouded in silk, and suffered injuries that led to his death in hospital on 23rd October 1935.

The sheer enjoyment normally brought about by Cobham's NAD tours was also offset by accidents caused sometimes by simple overconfidence and carelessness. Tyson, flying Avro Tutor G-ABZP, continued to steal the aerobatic limelight with inverted passes over the crowd enclosure at a height of twenty feet. He did however, make a slight concession to safety when he raised his altitude to some fifty feet when flying the length of Carlisle High Street, doing slow rolls all the way. At a subsequent court hearing, he pleaded the stunt was done in the interests of publicity, and the

judge was sufficiently impressed to impose only a nominal fine. But the following day, 30th July 1935, at Upwood Aerodrome, Ramsey, Huntingdonshire, George Eric Collins, chief flying instructor of the London Gliding Club and a very able sailplane pilot decided, against Tyson's advice, to attempt a 'bunt' or outside loop in a Grunau Baby glider. This aircraft was not designed to undertake such an extreme manoeuvre and the hideous crack that was heard as one wing detached itself, left no-one on the ground in doubt as to the inevitable and fatal outcome.

An incident that, remarkably, did not have a more serious outcome occurred when on 9th August, the so-called 'Ferry' show visited Llanelli, in Wales. With, as usual, no formal air traffic control it was incumbent on all pilots to ensure visually that clear paths were available for take-offs and landings. With a variety of aircraft milling about in such a relatively small airspace, this posed its own difficulties. On this particular occasion, the Airspeed Ferry's pilot, Joe King, found his view of the field partially blocked by farm buildings. He therefore failed to notice that Carruthers had aborted a take-off to allow another machine to lumber past in a different direction. The net result was that King's Ferry came round the corner of the buildings and cannoned into Carruther's Westland Wessex, both aircraft spinning round like a pair of locked scorpions. Dallas Eskell, having broken into the Ferry's fuselage with an axe to find that no-one was injured, then paraded the eight passengers on the grass. Addressing them quietly he said "I know you will feel bitterly disappointed at being deprived of your flight by this unfortunate incident, but if you will follow me, another pilot is waiting to take you up two at a time in the Avro". Apparently all followed as if mesmerised except for one little man who, firmly replacing his bowler hat exclaimed, "Not bloody likely" and promptly demanded a refund.

Prior to having divided the tour into two separate components, Cobham had taken the full entourage on another and this time 'farewell' visit to Ireland. To everyone's relief all went well without any of the potentially serious confrontations that had marred the previous excursions. It was at Tralee, however, that an amusing incident took place – one that Cobham for years afterwards called upon to illustrate the initiative frequently demonstrated by his NAD staff.

The smart trailer which carried Joan Meakin's glider; one of the more unusual events in the 1935 NAD tour.

Excessive heat from the engine exhaust systems had resulted in scorching of the back faces of the *"Astra's"* propellers. The problem required Cobham to get in touch with Bill Hartman the chief rigger and repair man, who had earlier returned to Ford. Asked if he could find a pair of replacements, he quickly tracked down two new four-bladed propellers and sent a cable informing Cobham that he was returning with them by rail. When, however, he arrived by lorry at Paddington, the railway staff were astounded that he should even consider it possible to transport them on a passenger train.

But Hartman produced his ground engineer's licence and forcibly explained that when travelling about the country on his various service jobs, he always took a pair of propellers in addition to his toolbox. Accordingly the station-master directed that a special open wagon be hitched onto the back of the Fishguard Mail Express. When he arrived in Ireland, a message awaited Hartman directing him to Tralee. With the confidence gained from his performance at Paddington, he successfully did the same thing again, thus arriving complete with 'personal baggage' only twenty-four hours after receiving Cobham's request for assistance.

Just over three weeks before the end of the 1935 tour, a disaster occurred before the largest crowd it was possible to imagine at that time. As described earlier, the 'Circus' always announced its arrival at a new location with a Grand Formation Flight. This feature had been impressively and safely performed over the preceding years at over 1,200 displays, but on Saturday 7th September all went catastrophically wrong. Leonard Carruthers, flying the Wessex G-ADFZ, was waiting at mid-day at 2,000 feet over the centre of Blackpool for the other aircraft to take up their normal positions alongside. Suddenly, without any warning, his machine was jolted upwards by a violent impact from below. The colliding aircraft, Avro 504N G-ACOD, flown by

Sir Alan Cobham with record breakers Jim Mollison and Amy Johnson at East Leys Farm, Grindale, near Bridlington.

Flt Lt Leonard Carruthers was fortunate to survive the mid-air collision over Blackpool on 9th August 1935 between Wessex G-ADFZ and Avro 504N G-ACOD.

Captain Hugh Stewart was instantly sliced in half. With fragments of wood and fabric entering the Wessex's cockpit and the 504's rudder having penetrated the fuselage, Carruthers fully expected his own aircraft to break up at any moment.

Unaware of the full extent of the accident, he carefully nursed the badly damaged machine back to Squires Gate to unload his complement of very fortunate, if bewildered, joy-riders. However, Stewart and his passengers, two young sisters, Lillian and Doris Barnes, all died with one of the girls being thrown out of the aircraft in full view of the holidaymakers jamming the sea front.

As to be expected, the event brought forth much comment in both the local and national press. But not all of it was adverse, as witnessed by the following which appeared in the *"Opinion"* column of the next day's *"Daily Express"*.

> "It is the sea which has made Great Britain great. It is the air which will maintain our greatness. To hold our place we must be predominant in the air, just as we have dominated the sea for generations. That is where Cobham's 'air circus' comes in. Its value as a national institution is that it is teaching people to be air minded. He is giving innumerable individuals their first venture off land and sea and after his instruction many of them become regular passengers of the great airline companies.
>
> "Now Cobham has suffered his first passenger loss *[sic]* since he embarked on his circus. If an omnibus company has a casualty we hear no criticism; no suggestion that the omnibus line should be closed down and go out of business. How much more valuable is Cobham's air display team than any omnibus line running on the ground."

Somewhat stilted prose perhaps, but it did establish a point of firm support by Lord Beaverbrook for Cobham as his NAD organisation struggled toward the end of its existence.

On 30th September 1935, all the touring machines were re-united at Dorking before finally leaving for the home depot at Ford. Soon afterwards, Cobham sold his organisation, but without the right to use his name, to Percy Phillips and Charles WA Scott who, throughout the summer, had often given demonstration flights and special commentaries at NAD events. Scott incidentally was not the only celebrity who made 'guest' appearances at Cobham's air shows; frequently attending with Puss Moth G-ABXY *"The Hearts Content"* were Jim Mollison and his wife Amy Johnson, who became firm friends of Cobham after Amy's flight in 1930 followed his own to Australia just four years earlier. In later years, Cobham made little secret of his great disappointment when, in contrast to his own earlier refusal to take on Jean Batten, Amy Johnson was unable to take up his offer to join his last tour.

Most of the National Aviation Display aircraft were included in the sale to form what, on 17th December 1935, became CWA Scott's Flying Displays Limited. A notable exception was the HP Clive *"Astra"* in which Hugh Johnson had carried some 250,000 passengers. Having been judged to be barely airworthy during the final shows, it had been nervously flown by Johnson, low and slow, back to Ford where it ended its days rather ignominiously, being burnt for the benefit of a motion picture film crew.

Scott's entourage went on the following year to tread what had now become a well-worn path around the country. His tour of Ireland was, however more exhaustive than those undertaken by Cobham and Barker-McEwan King. But, whatever success Scott achieved, it was insufficient to offset the appointment of a Receiver on 23rd September 1936 and just several weeks later in November, Scott's company was forced into liquidation.

As the year 1935 progressed, Cobham became less involved with public enlightenment and increasingly absorbed in starting up an airline, Cobham Air Routes Ltd, to operate between the mainland and Guernsey. He also found his time and efforts increasingly directed towards duties at Airspeed Ltd and other business concerns, and, most importantly, pursuing the art and science of refuelling aircraft in flight.

A poor but historic photo showing the end of the HP Clive, being cremated at Ford.

Times were indeed changing for many others who had played such major roles during the touring display years. Earlier in 1935, on 8th April, Captain Percy Phillips whose entrepreneurial instincts matched closely those of Sir Alan Cobham had, with H V 'Jock' Black, formed Air Publicity Ltd. The aim of the new company was to develop methods of aerial advertising that had previously been investigated by Cornwall Aviation Company and AJ Adam's Aeroplane Services Ltd.

The most popular form of message display involved the use of banners towed behind the wings or fuselage of one or more aircraft. An obvious choice for this work was the Avro 504, but at the end of the year, the choice of variant became limited when the Air Ministry refused to renew Certificates of Airworthiness for rotary-engined machines. It was therefore the Avro 504N, powered by the Armstrong Siddeley Mongoose or Lynx reciprocating piston engine that flew Air Publicity's banner. In the mid to late 1930s, Phillips' fleet consisted of eight Avro 504Ns, G-ACRE, G-ADET, 'DA, 'EV, 'BM, 'BO, 'BP and 'BR, two Airspeed Ferries, G-ABSI and G-ACFB, along with his long serving Spartan three-seater, G-ACAF.

With its administrative headquarters at London's Shell Mex House and its aircraft based at Heston, the new venture proved to

CWA Scott took over the Cobham operation for a single season in 1936; this scene taken at Hook aerodrome, near Kingston-upon-Thames on 16th April 1936 also shows an Airspeed Ferry on final approach.

Now wearing more stylish white overalls, recognisable in this 1935 picture are [from fourth left] Tresillian Bullmore, Cecil Bebb, Joe King, Hugh Johnson, Dallas Eskell and Geoffrey Tyson. Sitting on the ground are parachutists Naomi Heron-Maxwell and Ivor Price.

The Cobham tours were supplanted by his growing interest in air-to-air refuelling and his use of the former display airliners as tankers. Here, HP W.10 G-EBMR refuels Airspeed Courier G-ABXN over the Sussex coast.

Air Publicity was formed out of Cornwall Aviation Company and used many of the aircraft formerly on the Cobham and Scott tours including Avro 504N G-ACRE, in which owner Capt Phillips was later to lose his life.

be very successful. Although competing with Aerial Sites Ltd, Plane Advertising Ltd, and Publicity Planes Ltd, it was estimated that some 75% of the aerial advertising carried out over Britain at this time was conducted by an Air Publicity team that included ex-display pilots Mrs Winifred Crossley, Captain Morris E Hearn (not related to Martin Hearn), L Stonehill and JPG Bishop. Though profitable, 'PP' was not enamoured of this type of flying and considered it to be one of the most boring jobs in aviation. His extrovert personality found more expression in the aerobatic exhibitions he still performed whenever possible at air displays up and down the country. Perhaps inevitably it was this adventurous element that brought about his downfall on 13th February 1938.

Whilst flying in Avro 504N G-ACRE alongside Mrs Crossley's car heading towards her home, and with his wing tip almost touching the car window, all seemingly went well as 'PP' 'hedge-hopped' along the country lanes. But a manoeuvre which involved holding back to do a spectacular zoom up over the trees proved fatal. Misjudging the strength of a down current on

the far side of the copse, he crashed into a tree near her home at Gamlingay in Cambridgeshire. Although dragged from his burning aircraft and rushed by the Crossleys to Bedford Hospital, Phillips' injuries proved too severe and he died before reaching the operating theatre.

Air Publicity carried on with HV Black as managing director and Captain Hearn as chief pilot. At the time of the Munich crisis in September 1938, the company was able to offer the Air Ministry a package of pilot instructors, licensed ground engineers and aircraft that constituted a self contained training unit. However, as the immediate political tension eased, the offer was not taken up. But the following year, several of the company's Avros were requisitioned and allocated RAF serial numbers to be used by the RAF's Special Duties Flight to tow gliders out to sea from Swanage, on the Dorset coast. The trials successfully confirmed the ability of Britain's rudimentary coastal radar to detect wooden aircraft ie gliders, then thought likely to form part of Germany's invasion forces. This was the only occasion that RAF aircraft with a type-lineage directly related to the First World War took part in Second World War operations.

The dissolving of the National Aviation Day Display saw Geoffrey Tyson depart to become a test pilot at Avro, Cecil Bebb to join Olley Air Service and Leonard Carruthers rejoining the Service to become part of the instructing staff at the RAF's new flying school at Desford. With the worst of the Depression years now past, employment was relatively easy to come by, certainly for those skilled individuals who found that the qualification of having worked on the display tours virtually guaranteed them entry into any aircraft company.

A small number of Cobham's key NAD employees elected to stay with the restless, energetic entrepreneur, most notably Hugh Johnson, the ever versatile Percy Allison and Chris Tonge, who was destined to become Cobham's future 'money-man' and company secretary, together with long-serving secretary Frances Cameron. They then formed the nucleus of Flight Refuelling Limited, the company destined to become the cornerstone of what is today, Cobham plc.

Cobham himself now embarked on a new crusade, that of convincing civil and military authorities that air-to-air refuelling was about to transform long distance air travel. New challenges clearly beckoned, but that is a different story told fully elsewhere.

And finally . . . between-the-wars 'information technology'!

Cobham comes to town! The HP Clive leading the 'Grand Formation Flight' to announce the arrival of the National Aviation Day display.

Postscript...

A present day observer might well ask what this monumental effort to promote so many air displays achieved. The financial rewards were often tenuous, if not completely absent and it would appear that an unacceptable degree of danger and a correspondingly high accident rate, attended the shows. Some reflection on the laws governing the business as it evolved would, however, suggest how remarkable it was that the number of mishaps was not considerably higher.

Certainly Sir Alan Cobham's passionate aim of pioneering a network of civil aerodromes across the country to make 'Britain's Skyways, Britain's Highways' and imbuing the public with a need to fly was ahead of its time. The pace of Britain's aeronautical progress was also subject to other major influences both at home and overseas, and it was to be another quarter of a century, including the development accelerated by the war years, before the aeroplane became an affordable and available means of travel for the general population.

What the displays did do, however, in addition to creating some hundreds of much needed jobs, was to create an enormous source of pleasure and fun for so many ordinary men and women who often suffered extreme hardship during the years of the Depression. Cobham estimated that some three million people paid to see his shows with probably an equal number that did not.

Records also showed that 990,000 people took to the skies in his aircraft for a thrill that they would frequently and fondly recall for the rest of their lives.

But, beyond these simple statistics, who could put a value on the sense of adventure such an experience created in the minds of those impressionable youngsters who not only took up careers within an exciting and expanding industry, but more visibly, were soon to be called upon to serve in the air in the forthcoming conflict?

The National Aviation Display's announcer Tresillian Bulmore, later became a member of the Royal Air Force's selection panel. He observed that, when asked if they had ever flown before, some 75% of the aircrew candidates that came forward in 1939-40, proudly replied "Yes – with Cobham's Flying Circus"! This fine and timely tribute was one which Sir Alan was particularly pleased to recall up until his death, aged seventy-nine, on 21st October, 1973.

The contribution made by Cobham and so many others to Britain's awareness of 'air matters' cannot be measured in finite terms. But its influence, perhaps thought to have disappeared with the blue haze of castor oil-fuelled exhausts, is still surely evident whenever a present day airshow commentator announces, "Ladies and Gentlemen, the aircraft now approaching from your left..........!"

A flypast led by the ubiquitous Airspeed Ferry opens the NAD show at Biggin Hill on 20th May 1932.

APPENDIX I: The Mandate

The text of Alan Cobham's 1932 Mandate was as follows:

The Royal Aeronautical Society

That as a nation's progress in aviation is determined by the technical knowledge at the disposal of its aircraft constructors, it is incumbent upon Great Britain to safeguard the leadership which has been won in this field by placing research at the forefront of every policy of development, for it is only by research that advancement can be achieved.

The Royal Aero Club

That air racing and record breaking achievements should be given greater encouragement by the Government and the fact recognised that there is no sounder investment for the British nation than to build aircraft for record breaking and air racing because –

(a) The scientific knowledge gained from such achievements has done more to improve British aircraft than possibly any other single activity. The record speed of today is the commercial speed of tomorrow.

(b) The sporting nature of flying has always been the backbone of British interest in the air, and continued endeavour in this direction is necessary to the life of British aviation.

The Air League of the British Empire

Believing that Britain's future depends upon a closer linking up of the Empire and that this can best be done by means of aviation, it is, in our opinion, essential:

(1) To increase the frequency and speed of existing air routes.

(2) To extend the Indian air route to Australia.

(3) To explore air routes and to establish services with Canada and in the West Indies.

(4) To establish twenty-four hour services of mails.

(5) To bring home to the public the vital importance of aviation to the security and well-being of the Empire.

The Guild of Air Pilots and Air Navigators

That the competence of civil and commercial air pilots is a matter of vital importance to the general public and that the following recommendations of the Guild of Air Pilots and Air Navigators should be adopted forthwith:

(a) That no persons should be employed as instructors in flying unless their qualifications comply with standards adopted by the Guild.

(b) That the standard of training of commercial air pilots in air navigation should progressively be raised to higher levels.

General Council of Associated Light Aeroplane Clubs

That results of great national importance have been accomplished by the light aeroplane clubs and that it behoves the nation to give the fullest support to these clubs to enable them to carry on and extend their good work. The public should grant their forbearance for any inconvenience that flying may cause, because this inconvenience will disappear with the improvement of aircraft. Municipalities should give every assistance and encouragement to their local clubs, and the Government should continue to aid the movement until such time as every club is firmly established.

The British Gliding Association

That the Government make a grant of an adequate sum of money spread over a number of years, (a) for the establishment and maintenance of Central Scientific and Training Stations to enable the Youth of Britain, and scientifically and technically minded people, to develop and carry out research work through motorless flying for the benefit of British aviation; (b) a grant be made for the establishment of a central fund available for the purpose of assisting clubs to establish flying grounds, erect hangars, buy machines and equipment. Then the British Gliding Association be entrusted with the establishment, maintenance, direction and control of such stations and the administration of the grants to clubs.

The Association of British Chambers of Commerce

That in view of the rapid developments which are taking place in aviation, suitable sites for aerodromes should be secured in the vicinities of the principal centres and allocated to that purpose without delay.

The London Chamber of Commerce Civil Aviation Section

That, as British aviation is one of the most important factors in the development of trade:

(a) Civil and commercial aviation in all its aspects should be developed to the fullest extent.

(b) Aviation activities should be speedily inaugurated in all parts of the British Isles and

(c) Every support should be given to the plea of the Association of British Chambers of Commerce that municipal authorities should foster the developments of aerodromes, thus making it possible for every town to have facilities for the reception of aircraft.

The Automobile Association

That owners of suitable sites be urged to give through the AA temporary permission for aeroplanes to land in case of need; civil authorities are asked to co-operate in the service of weather reports and forecasts now being broadcast by the AA from Heston, and to arrange for the reception and recording of the broadcasts so that they may always be available to enquirers. By these means the development of British air travel will be hastened.

The Society of Model Aeronautical Engineers

That the model aircraft societies and clubs should be encouraged by educational authorities throughout the country with the ultimate object of establishing classes for the study of aeronautics and of forming model aeroplane clubs at each school for the practical demonstration of model work.

The Model Aircraft Club (TMAC)

That municipal authorities should do all in their power to assist model aeroplane clubs and, in particular, should give facilities on open spaces for the clubs to conduct their meetings.

APPENDIX II
Joy-Riding and Flying Circus Operators

Listed in this Appendix are known details of all the operators of joy-riding and flying circus businesses in the British Isles in the inter-war period. It excludes clubs and other commercial concerns where such activities were simply peripheral to their main business. Also not included are aircraft which were not part of the joy-riding or circus activities.

Certain abbreviations have been used for convenience:

Appt	Date Appointed		GE	Ground Engineer
Bt	Bought		Mgr	Manager
Cld	Cancelled		Op	Operated
CofA	Certificate of Airworthiness		Regd	Date Registered
Dbf	Destroyed by fire		Regn	Registration
Dd	Delivered		Wfu	Withdrawn from use
Fndr/dirs	Founder/directors			

AERIAL PHOTOS LTD
81A George St, Edinburgh
Formed 18.7.19, operated a pleasure flight tour of St Andrews, Rothesay, Dundee, Perth, Berwick & Montrose in 1920; carrying 1,509 passengers
Dirs: Capt RSJB Andrews, Major CH Chichester-Smith, Capt O Hardie [possibly also Capt Archibald N Kingwill 1919?]
Pilots – Edward A Jones [appt chief pilot 8.19]; Capt Simpson & Lt Rendle Woodcock [9.19]; Archibald Kingwill [4.20]
Investors – HE Haig [papermaker]; AR Wilson Wood [gentleman]

Avro 504K	G-EAHU	Regd 7.19; Crashed Turnhouse 28.4.22 and sold WL Woodward 5.22
Avro 504K	G-EAHV	Regd 3.9.19; CofA lapsed 7.8.21
Avro 504K	G-EAIG	Regd 3.9.19; No CofA; regn cld 10.19
Avro 504K	G-EALE	Regd 8.19; sold Midland Aviation Co 8.22
Avro 504K	G-EAGB	Regd 12.19; sold Midland Aviation Co 8.22
Avro 504K	G-EAGC	Regd 12.19; CofA lapsed 9.11.22; sold Manchester Aviation Co Ltd 3.23

AEROPLANE SERVICES LTD
1a Park Road, Leyton, London E10; to 156 Edmund St, Birmingham
Operated from The Flying Ground, Holden Rd, North Finchley; later to Castle Bromwich. Summer 1929 season at Southend in competition with Surrey Flying Services
Formed 27.3.29 by Albert J Adams, CC Bint, CH Bint
Receiver appointed 6.12.29

Avro 504K	G-AAEZ	Regd 3.29; sold HC Howard 5.30

AIRCRAFT DEMONSTRATIONS LTD
West India House, 101 Leadenhall St, London EC3 [based Croydon]
Formed 28.11.36 by Harry Barker & Jimmy McEwen King; commenced trading 4.37
Formal directors were Frederick MacKenzie MacGregor & Sidney R Vennervald [both nominees]
Operated as **Coronation Air Displays** [tour commenced 4.37 [33 towns]; then to Irish Free State until 8.37]
Pilots included Lance Rimmer; Gerry Chambers & Richard Fotheringay Robinson; Parachutist Harry Ward
Operated 3 unidentified aircraft; assumed Avro 504s, bought 2.37. One crashed Kinsale; the second was taken to store at Croydon; nothing known on the third.
Company into **liquidation** 11.37

AIR PAGEANTS LTD
66 Fenchurch St, London EC3 [based Croydon]
Formed 23.6.32 by Henry ACG Barker & James McEwen King
Operated as **British Hospitals Air Pageants** 1933 & 1934; **Sky Devils Air Circus 1934**
Receiver appointed 3.10.34 and reformed as **Jubilee Air Displays** 1935

Avro 621 Trainer	G-ACOV	Regd 21.3.34; Crashed Stretton, nr Burton-on-Trent 17.6.34
Miles Hawk	G-ACPC	Regd 26.3.34; CofA lapsed 2.4.35
DH.60M	G-ACOA	Regd 4.34; sold Aircraft Exchange & Mart Ltd 7.35
Avro 504N	G-ACNV	Regd 24.4.34; CofA lapsed 10.4.35

AIR TOURS LTD
20 Market St, Mayfair, London W
Formed 15.3.34 – successor to Aviation Tours Ltd
Fndr/dirs: Earl B Fielden [chmn/MD]; Capt Arthur D Makins [of Brighton Road Racing Ltd, 20 Market St]; Samuel L Harris [of Fasta Films Ltd]

Operated with Cobham/National Aviation Day Ltd

AW Argosy	G-EBLF	Hired from Imperial Airways Ltd 4.34-10.34

AIR TRANSPORT & SALES LTD
Hayling Aerodrome, North Hayling, Hants; also operated from fields at Eastoke, Hayling & Horndean
Formed as **Air Transport Company** 4.32 by George Morgan-Harris & Denis I Peacock operating 2 ex-Skywork Spartan 3-seaters
UK display tour commenced 4.32, incl John Tranum [parachutist] & Bud Cummings [pilot]
Coy formed 30.8.33 [but principally to operate West Malling Airfield]

Spartan 3-str	G-ABRA	Op 4.32; sold J Stark 10.32
Spartan 3-str	G-ABRB	Op 4.32; sold as VR-TAJ 10.32
Spartan 3-str	unknown	Op at Hayling 8.33
Bluebird	G-AAUX?	Op at Hayling 8.33
Spartan 3-str	G-ABYH	Regd 11.6.34; Fatal crash Hayling Island 20.7.35
Spartan 3-str II	G-ABTR	Regd 21.1.37; sold FG Barnard 4.37

AIR TRAVEL LTD
Bridge Buildings, Wisbech, Cambs; to Penshurst Aerodrome, Kent 1933; to London South [Gatwick] Airport, Horley, Surrey 1.34 [also operated from Walsall 11.36]
Formed 21.3.32 by Fred JV Holmes & Capt Archibald N Kingwill
Pilots: Archibald Kingwill, W 'Dusty' Miller, George Kemp [1932/36] operated as *"Silver Trio"*
Business and assets bought by **Airports Ltd** 9.37 [Holmes and Kingwill were retained as Works & Sales mgrs respectively]

Avro 504K	G-ABUK	Regd FJV Holmes 1.3.32; Crashed Hedon, Hull 4.32
Avro 504K	G-ABVH	Regd FJV Holmes 3.32; regd to coy 12.1.34; CofA lapsed 1.8.36
Avro 504K/N	G-ABVY	Regd FJV Holmes 4.32; regd to coy 12.1.34; CofA lapsed 8.3.35
Avro 504K/N	G-ACCX	Regd FJV Holmes 8.2.33; Crashed Stewkley, Bucks 19.6.34
Avro 504K	G-ABSN	Regd 11.8.33; sold FB Chapman 12.34
Avro 504N	G-ACLV	Regd 11.33; sold National Aviation Displays Ltd 3.35
Avro 504N	G-ACOD	Regd 3.34; sold National Aviation Displays Ltd 3.35
Avro 504N	G-ACOM	Regd 3.34; sold Plane Advertising Ltd 6.37
Avro 504N	G-ADBD	Regd 1.35; sold National Aviation Displays Ltd 6.35; *see below*
Avro 504N	G-ADBM	Regd 1.35; sold HC Wardle 3.35
Avro 504N	G-ADBD	Regd 6.11.35; sold Canute Air Park Co 8.36
Avro 504N	G-AEDD	Regd 4.36; sold Plane Advertising Ltd 5.36
Avro 504N	G-AEGW	Regd 22.4.36; sold LJ Rimmer/WF Davison 3.37
Avro 504N	G-AEIJ	Regd 5.36; sold Plane Advertising Ltd 7.36
Avro 504N	G-AEMP	Regd 8.36; sold Plane Advertising Ltd 9.36

AIR TRIPS LTD
Boughton House, 10/12 Church Rd, Tunbridge Wells, Kent [based Croydon]
Coy formed 27.3.34 by Pauline Gower and Dorothy Spicer as successor to trading partnership which had commenced 1931 from a field nr Wallingford, Berks
Based Hunstanton for 1934 season [with parachutist Bruce 'Bill' Williams]
Flew with Crimson Fleet [Modern Airways] 1932; British Hospitals Air Pageant 1933; Jubilee Air Displays 1935; British Empire Air Displays 1936
Ceased trading 6.37 – having carried over 25,000 passengers

Spartan 3-str	G-ABKK	Regd P Gower 1.32; regd to coy 9.11.34; named *"Helen of Troy"*; Collided with Avro 504N G-AECR & crashed Westwood, Coventry 10.5.36
DH.83	G-ADNF	Op 8.35 [Regd to CT Berry]; Crashed Hunstanton 11.8.35
Spartan 3-str	G-ACAD	Regd 15.7.36; sold DB Prentice 8.37

ALAN COBHAM AVIATION LTD
150 New Bond St, London W1; to Grand Buildings, Trafalgar Sq, London WC2 [11.29] [based Croydon/Ford]

Formed 22.11.26 initially as joint venture with Warwick Wright Ltd; directors – Sir Alan Cobham; Emil A Merckel [of Warwick Wright Ltd] and Lt Col Warwick Wright. Became wholly-owned by Cobham 4.30.

Owned two aircraft which were used for joy-riding etc – see National Aviation Day Ltd etc

DH.61	G-AAEV	Regd 2.29 *"Youth of Britain"*; sold Imperial Airways Ltd 1.30
HP Clive	G-ABYX	Regd 10.4.33; sold National Aviation Displays Ltd 11.33

ALPHA FLYING SERVICE
13 The Ridgeway, Enfield

Prop: Major Edward G Clerk

Operated from Rye, Sussex 1934-35; and reportedly from Weymouth with Avro 504N G-ACRS

Spartan 3-str	G-ABJS	Regd EG Clerk 30.10.34; sold as VH-UUU 1935

ANDERSON & POOL AVIATION CO LTD
Avoca, Farnborough, Kent [based Priory Heath, Ipswich]

Prop: Capt Anderson

Avro 504K	G-EAIB	Regd 21.6.20; dd 2.7.20; CofA lapsed 7.8.20

AVIATION DEVELOPMENTS LTD
100 Charing Cross Rd, London WC2

Formed 1.12.33 by Ronald Dixie-Gerrans, Sydney N Gerrans & Richard L Howard-Flanders

"Flying Fair" Tour commenced 31.3.34 at Southend/Rochford; planned to visit 150 towns in Southern England; probably abandoned during April 1934

Venture also involved WB Grieve & Eric A Starling

Redwing II	G-ABOK	Hired from Redwing Aircraft Co Ltd 3.34; sold GJ Dawson 6.34

AVIATION TOURS LTD
Stafford Rd, Wallington, Surrey [based Croydon]; 26 Swadford St, Skipton, Yorks [6.32] [based Hanworth]; Grand Buildings, Trafalgar Square, WC2 [6.32]; 66 Fenchurch St, London EC3 [4.33] [based Ford]; 20 Market St, Mayfair, London W1 [1935]

Formed 3.30 by Capt Earl B Fielden [MD & chief pilot] and Frank WJ Grant [of Surrey Flying Services]

Pilots incl Harold Lawson; J Wilson & EW Jordan; Wingwalker – Harry Willis

Refinanced by Sir Alan Cobham 1932 and flew for National Aviation Day Ltd 1932

Flew for British Hospitals Air Pageants 1933

Bought out Western Aviation Ltd 1933

Ceased trading and reformed as Air Tours Ltd 3.34

Avro 504K	G-EBYW	Regd 5.30; sold EG Clerk 6.34
HP W8B	G-EBBI	*"Prince Henry"* – Hired from Imperial Airways 4.30-9.30; 4.31-9.31 [reportedly operated for 1932 season but unlikely since CofA lapsed 24.3.32]
HP W.10	G-EBMR	*"City of Pretoria"* – Hired from Imperial Airways 4.32 & sold Sir Alan Cobham 3.33
DH.83	G-ABUP	Regd 6.6.32; crashed Ashby, nr Scunthorpe 24.8.33
HP W.8E	G-ACDO	Bought 2.33 but not overhauled and used for spares 1934
Avro 504K	G-EBQR	Regd 11.4.33; Not used since CofA lapsed 26.6.31
Avro 504K	G-EBXV	Regd 11.4.33; Not used since CofA lapsed 22.3.33

AVRO TRANSPORT CO
Operation formed by AV Roe & Co Ltd 1919; ceased trading 1920/21

Initial operations wef Easter 1919 at Hamble [pilots GLP Henderson, HA Hamersley, F Warren-Merriam] and Alexandra Park [opened 6.19]; also Blackpool Sands [opened 5.19]

Expanded 6.19 under overall control of GLP Henderson with operations at:
Hounslow Heath [Major AG Taylor]
Manston/Margate [Capt Duncan Davis]
Southsea/Eastney [Flt Lt EA Sullock]
Weston-super-Mare [Capt Denis G Westgarth-Heslam]
Ladies Mile, Brighton/Patcham/Blatchington Farm [Capt Donald IM Kennard]
Swansea/Brynmill [Capt FGM Sparks]

Avro 504K G-EBYW of Aviation Tours with wing-walker Harry Willis.

Avro 504K E4359 (G-EABJ) with another on Blackpool beach.

Porthcawl [Capt Donald Herne]
Blackpool/South [Maj McMinnies]
Southport/Birkdale Sands [Capt BR Collison]
Fleetwood/Scale Hall, Lancaster/Morecambe Sands [Lt JF MacRae]
Rhyl [Capt E Maitland Heriot]
Liverpool/Waterloo Sands/Manchester/Alexandra Park [GLP Henderson]
Douglas Promenade/Ramsay, IoM [Lt GB Moxon]
Also in Netherlands 7.19-1.20 [under command of Capt Walter GR
   Hinchliffe – pilots Lt Brown, Lt Shanks, Lt Roberts, Lt Inglis]
The following were probably operated by Avro Transport Company during
the 1919 season; it is possible that some of the later ones were held in
store pending issue but were not so used. For the 1920 season, some of
the aircraft would have been hired out to pilots for joy-riding purposes.

Avro 504K	G-EAAK	Regd 30.4.19 [E4222] [based Hounslow Heath]; CofA lapsed 8.5.20
Avro 504K	G-EAAL	Regd 30.4.19 [E4154] [based Southsea]; Wfu and converted to Avro 548 2.21
Avro 504K	G-EAAM	Regd 30.4.19 [E3289]; [based Hounslow Heath]; CofA lapsed 8.5.20
Avro 504K	G-EAAN	Regd 30.4.19 [E4225]; [based Weston-super-Mare]; Crashed Green Common, Weston-super-Mare 28.8.19 [as E4225]
Avro 536	G-EAAO	ATC.4; Regd 30.4.19 [K-106]; [based Hounslow Heath]; regn cld 7.19
Avro 536	G-EAAP	ATC.3; Regd 30.4.19 [K-105]; [based Hounslow Heath]; Fatal crash on landing Sandhurst 6.8.19 [as K-105]
Avro 536	G-EAAQ	ATC.2; Regd 30.4.19 [K-104]; [based Hounslow Heath]; Crashed Southwark Park 9.9.19 [as K-104]
Avro 504K	G-EABJ	Regd 19.5.19 [E4359]; [based Blackpool][Fleet No.3]; CofA lapsed 18.5.20
Avro 504K	G-EABK	Regd 27.5.19 [D8287]; [based Southport]; CofA lapsed 19.5.20
Avro 504K	G-EABL	Regd 9.5.19 [E4324] [based Alexandra Park]; CofA lapsed 18.5.20
Avro 504K	G-EABM	Regd 9.5.19 [E4360]; [based Southport/IoM][Fleet No.2]; CofA lapsed 18.5.20
Avro 504K	G-EABV	Regd 22.5.19 [D6229]; [based Blackpool][Fleet No.5]; regn cld 12.19
Avro 504L/536	G-EACC	ATC.1; Regd 30.6.19 [K-114]; [based IoW]; CofA lapsed 30.6.20
Avro 504K	G-EACD	Regd 15.5.19 [E4224]; regn cld 12.19
Avro 536	G-EACG	ATC.9; Regd 20.5.19 [K-116]; [based Southsea]; Crashed Manston 30.8.19
Avro 538 Scout	G-EACR	Regd 29.5.19 [K-132]; [Fleet No.7 – Operated by JCC Taylor, ATC's chief engineer]; regn cld 9.20
Avro 504K	G-EACS	Regd 29.5.19 [D9341]; CofA lapsed 21.5.20
Avro 504K	G-EACW	Regd 29.5.19 [E1663]; [based Southport]; Crashed in sea off Blackpool 4.8.19
Avro 504M	G-EACX	ATC.10; Regd 29.5.19 [K-134]; [based Hounslow Heath]; CofA lapsed 23.5.20
Avro 504K	G-EADA	Regd 29.5.19 [E4221]; CofA lapsed 20.5.21; [later used as Avro 504N/Bristol Lucifer development aircraft]
Avro 536	G-EADC	ATC.5; Regd 6.6.19 [K-137]; [based Southsea]; CofA lapsed 4.6.20
Avro 504K	G-EADD	Regd 6.6.19 [E1665]; [based Manston]; CofA lapsed 30.5.20
Avro 504K	G-EADI	Regd 5.6.19 [E3292]; [based Morecambe] [Fleet No 6]; CofA lapsed 26.5.20
Avro 504K/L	G-EADJ	Regd 5.6.19 [H2581]; [based Cockshott, Lake Windermere]; CofA lapsed 25.6.20
Avro 504K/L	G-EADK	Regd 2.8.19 [H2582]; [based Cockshott. Lake Windermere; later to Folkestone]; CofA lapsed 17.7.21
Avro 504K	G-EADM	Regd 5.6.19 [E4336]; [based Southsea]; sold to Belgian AF 4.21
Avro 504K	G-EADN	Regd 5.6.19 [E3293]; sold to Belgian AF 4.21
Avro 504K	G-EADW	Regd 6.6.19 [E4343]; Damaged in gales Rhyl 27.8.19
Avro 504K	G-EADX	Regd 6.6.19 [D6239]; [based Southport] [Fleet No.9]; CofA lapsed 25.5.20
Avro 504K	G-EAEN	Regd 11.6.19 [D9018]; [based Southport] [Fleet No.11]; regn cld 1/20
Avro 504K	G-EAEO	Regd 11.6.19 [E4362]; [based Morecambe] [Fleet No.10]; CofA lapsed 14.6.20
Avro 504K	G-EAEV	Regd 18.6.19 [H2586]; CofA lapsed 11.6.20
Avro 504K	G-EAFC	Regd 19.6.19 [E1660]; [based Rhyl] [Fleet No.16]; Damaged in gales Rhyl 8.19
Avro 504K	G-EAFD	Regd 19.6.19 ["D4329"/E4329]; [based Southport] [Fleet No.12]; sold J Blake & Co 3/20
Avro 504K	G-EAFE	Regd 19.6.19 [D7648]; CofA lapsed 2.7.21
Avro 504L	G-EAFF	ATC.12; Regd 19.6.19 [K-145]; [based Hayling Is/IoW]; sold to Belgian AF 7.21
Avro 504L	G-EAFG	ATC.13; Regd 19.6.19 [K-146]; [based Paignton]; Crashed off Alderney 5.10.19
Avro 504K/548	G-EAFH	ATC.14; Regd 19.6.19 [K-147]; sold Welsh Avn Co Ltd 5.21
Avro 504K	G-EAFS	Regd 1.7.19 [D9340]; [based Southport] [Fleet No 15]; CofA lapsed 3.7.20
Avro 536	G-EAGM	ATC.6; Regd 10.7.19 [K-161]; [based Weston-super-Mare]; Crashed Weston-super-Mare 1.9.19 [as K-161]
Avro 504K	G-EAGO	Regd 12.7.19 [D9343]; [based Southport]; Crashed Southport Beach 20.8.19
Avro 504K/L	G-EAGU	Regd 14.7.19 [H2585]; CofA lapsed 15.8.20
Avro 536	G-EAHA	ATC.7; Regd 14.7.19 [K-165]; [based Amsterdam]; regn cld 9.20 – probably sold locally
Avro 536	G-EAHB	ATC.8; Regd 14.7.19 [K-166]; [based Margate]; CofA lapsed 14.7.20
Avro 536	G-EAID	Regd 31.7.19 [K-173]; CofA lapsed 29.7.20
Avro 536	G-EAIE	Regd 24.7.19 [K-174]; [based

		Blackpool][Fleet No 19]; CofA lapsed 29.7.20
Avro 536	G-EAIF	Regd 7.8.19 [K-175]; CofA lapsed 29.7.20
Avro 504K	G-EAIH	Regd 31.7.19; [based Amsterdam]; sold to Belgian AF 4.21
Avro 504K	G-EAII	Regd 31.7.19; Crashed Great Yarmouth 1.8.20
Avro 504K	G-EAIJ	Regd 31.7.19; [based Amsterdam]; CofA lapsed 29.7.20
Avro 504K	G-EAJQ	Regd 8.19; [based Manston]; sold HWB Hansford 8.20
Avro 536	G-EAJR	Regd 8.8.19; CofA lapsed 19.8.20
Avro 504L	G-EAJX	ATC.16; Regd 12.8.19; CofA lapsed 19.8.20
Avro 504K	G-EAJZ	Regd 8.19; [based West Blatchington, Hove]; sold WG Pudney 11.22
Avro 504K/L	G-EAKA	Regd 14.8.19; sold Chile 8.21
Avro 504K	G-EAKB	Regd 14.8.19; CofA lapsed 21.8.20
Avro 536	G-EAKD	Regd 16.8.19; CofA lapsed 21.8.20
Avro 536	G-EAKJ	Regd 8.19; [based Brighton]; CofA lapsed 2.9.21; later sold Surrey Flying Services Ltd 4.24
Avro 536	G-EAKK	Regd 18.8.19; CofA lapsed 4.9.20
Avro 536	G-EAKL	Regd 18.8.19; CofA lapsed 14.11.20
Avro 504K	G-EAKW	Regd 20.8.19; [based Hamble]; CofA lapsed 24.8.20
Avro 504K	G-EAKX	Regd 8.19; [based Brighton]; sold 8.20; to Berkshire Aviation Co 12.20
Avro 504K	G-EAKY	Regd 20.8.19; CofA lapsed 27.8.20
Avro 504K	G-EAKZ	Regd 20.8.19; [based Brighton]; CofA lapsed 7.9.20
Avro 504K	G-EALA	Regd 8.19; sold R Taylor 7.22
Avro 504K/L	G-EALB	Regd 20.8.19; CofA lapsed 15.6.20
Avro 504K/548	G-EALF	Regd 20.8.19; rebuilt as Avro 548 12.20
Avro 504L	G-EALH	ATC.15; Regd 21.8.19; CofA lapsed 24.8.20
Avro 504L	G-EALI	ATC.17; Regd 21.8.19; CofA lapsed 29.8.20
Avro 504K	G-EAMO	Regd 10.9.19; [based West Blatchington Hove]; CofA lapsed 22.9.20; sold to Chile 8.9.21
Avro 504K	G-EAMP	Regd 10.9.19; CofA lapsed 6.10.20
Avro 504K	G-EAMQ	Regd 10.9.19; Crashed Rhyl 14.10.20
Avro 504K	G-EAMZ	Regd 9.19; sold FP Raynham 5.23
Avro 504L	G-EANB	ATC.18; Regd 16.9.19; sold to Sweden as S-IAA 6.21
Avro 504K	G-EAND	Regd 18.9.19; [based West Blatchington, Hove]; CofA lapsed 24.9.20; sold to Belgian AF 4.21
Avro 504K	G-EANE	Regd 18.9.19; CofA lapsed 22.10.20; sold to Chile 8.9.21
Avro 504K	G-EANF	Regd 18.9.19; CofA lapsed 30.9.20
Avro 504K	G-EANG	Regd 18.9.19; CofA lapsed 4.10.20; sold to Belgian AF 4.21
Avro 504K	G-EANO	Regd 22.9.19; CofA lapsed 8.10.20
Avro 536/546	G-EAOM	ATC.23; Regd 14.10.19; [based Blatchington Farm]; CofA lapsed 16.12.21

BERKSHIRE AVIATION CO/BERKSHIRE AVIATION TOURS LTD
The Mulberries, East Hanney, Wantage, Berks [based East Hanney]

Formed as **BERKSHIRE AVIATION CO** 4.19 by Fred JV Holmes; John DV 'Jack' Holmes [resigned 1923] and Alan Cobham [resigned 1920]; Capt AL Robinson [joined 1.20; dep summer 1922 to Daimler Hire]; Joe CC Taylor [joined early 1920; departed pre 1924]; OP Jones [joined 1920; left 12.20]

Also traded under name of **Cobham-Holmes Aviation Co** 3.20

Reformed as **BERKSHIRE AVIATION CO LTD** 25.8.21 [to acquire business of JDV Holmes/JCC Taylor]

Adopted name **Berkshire Aviation Tours** [by 5.24] and reformed as **BERKSHIRE AVIATION TOURS LTD** 10.8.26

Other pilots: Sydney F Woods [12.22]; JD Parkinson [5.24; to senior pilot 1926; to Newcastle Aero Club 7.26]; Clifford S Kent [appt 5.24]; FGM Sparks [5.24]; Geoffrey R Beck [appt 11.25; left 4.27]; Lionel Leleu [11.25; to Imperial Airways 1926]; J Stirling [appt 5.26]; Lance Rimmer [10.26-12.27 & 1928]; Tommy Nash [12.27]; Harold Lawson [1928]; EE Fresson [3.28-9.28]; AV Heaton [9.29]

Also employed H Gomez-Cornejo & T Graham-Wolland as wing-walkers

Also based at Witney, Oxon and Wythenshaw before going to Barton

Became subsidiary of **Northern Air Transport Ltd** 4.29, and moved base to Barton Moss; but continued to operate under Berkshire name until Group went into receivership 18.11.31.

Most of the fleet were registered in individuals' names and latterly under Northern Air Transport name.

Avro 504K	G-EACL	Regd JDV Holmes 20.5.19 [D9298]; Crashed nr Rush Mills, Northampton 21.7.19; not repaired
Avro 504K	G-EAIB	Regd JDV Holmes 8.19]; [sold Anderson & Pool Avn Co 6.20
Avro 504K	G-EASF	Regd Cobham & Holmes Avn Co 3.20; regd to OP Jones 7.20; regd to FJV Holmes 3.22; regd Northern Air Transport Ltd 6.29; sold LJ Rimmer/W Mackay 7.33
Avro 504K	G-EAKX	Regd 12.20; [Fleet No 4]; [Operated in Air Taxis c/s 1921]; regd FJV Holmes 4.23; CofA lapsed 2.6.28; regd Northern Air Transport Ltd 6.29; sold LJ Rimmer/W Mackay 7.33
Avro 504K	G-EAHZ	Regd FJV Holmes 16.8.21; regn cld 1.3.22
Avro 504K	G-EBCK	Regd FJV Holmes 4.22; sold F Neale 1.23
Avro 504K	G-EBFV	Regd FJV Holmes 19.4.23; sold Wm Beardmore & Co Ltd 10.23
Avro 536	G-EAKN	Regd FJV Holmes 21.9.23; Crashed Dorton, nr Brill, Bucks 12.8.24
Avro 504K	G-EBIN	Regd FJV Holmes 28.11.23; Crashed Plymouth 25.7.24
Avro 504K	G-EBKB	Regd FJV Holmes 9.24; regd Northern Air Transport Ltd 18.4.29; Crashed off Scarborough 13.9.31
Avro 504K	G-EBKR	Regd FJV Holmes 12.24; sold CS Kent 8.27
Avro 504K	G-EBKX	Regd FJV Holmes 2.25; regd Northern Air Transport Ltd 7.29; sold LJ Rimmer/W Mackay 7.33
Avro 504K	G-EBOB	Regd FJV Holmes 10.5.26; CofA lapsed 3.3.29
DH.6	G-EBPN	Regd FJV Holmes 11.26; sold British Flying & Motor Services Ltd 4.28
Avro 504K	G-EBSL	Regd FJV Holmes 7.27; regd Northern Air Transport Ltd 10.5.30; CofA lapsed 22.2.32
Avro 504K	G-EBSM	Regd FJV Holmes 7.27; sold James Bunning Ltd 5.28
DH.6	G-EBVS	Regd FJV Holmes 1.28 [unconfirmed]; Not certified & sold British Flying & Motor Services Ltd 4/28
Avro 504K	G-EBVW	Regd FJV Holmes 1.28; regd Northern Air Transport Ltd 18.7.29; CofA lapsed 23.2.32; regn cld 12.33
Avro 504K	G-AAEZ	Regd HC Howard; Regd Northern Air Transport Ltd 8.31; sold LJ Rimmer/W Mackay 7.33
Avro 504K	G-AAYH	Regd Northern Air Transport Ltd 5.5.30; sold to Ireland as EI-AAM 3.32
Avro 504K	G-ABAV	Regd Northern Air Transport Ltd 2.6.30; CofA lapsed 8.7.31
Avro 504K	G-ABAW	Regd Northern Air Transport Ltd 2.6.30; CofA lapsed 14.7.32
Avro 504K	G-ABHI	Regd Northern Air Transport Ltd 9.12.30; CofA lapsed 24.5.33
Avro 504K	G-ABHJ	Regd Northern Air Transport Ltd 12.30; sold LJ Rimmer/W Mackay 7.33
Avro 504K	G-ABHK	Regd Northern Air Transport Ltd 12.30; sold LJ Rimmer/W Mackay 7.33
Avro 504K	G-ABHP	Regd Northern Air Transport Ltd 17.12.30; sold to Ireland as EI-AAN 3.32
Avro 504K	G-ABLL	Regd Northern Air Transport Ltd 4.31; sold LJ Rimmer/W Mackay 7.33

BLACKPOOL AIR SERVICE
2 Birley St, Blackpool; later 20 Central Drive, Blackpool [based Squires Gate]

Possibly also known as Blackpool Flying Services

DH.6	G-EAUS	Regd 2.5.21; CofA lapsed 2.7.23; – to Manchester Aviation Co for spares
DH.6	G-EAUT	Regd 2.5.21; CofA lapsed 7.5.22; regn cld 20.3.23

BORDER AVIATION CO LTD
**61 Botchergate, Carlisle and 11 Castle St, Carlisle [based Carlisle] –
also joyriding at Oliver Mount, Scarborough**
Formed 19.4.20 by Robert F [Bob] Little, Percy H Ingham, RJ Cairns; [later
joined by Graham Little – brother]
Pilot: Capt John Oliver
Ceased trading 10.20 [possibly into liquidation] and assets sold to Ingham
& Little Aviation Co.

Avro 504K	G-EANQ	Regd 4.20; sold Ingham & Little Avn Co 5.21
Avro 504K	G-EAIA	Regd 8.20; sold Ingham & Little Avn Co 6.21

BOURNEMOUTH AVIATION CO LTD
**Ensbury Park, Bournemouth [based Ensbury Park; aka Moordown;
but joy-riding also in Weymouth/Bath areas]**
Formed 5.19 by FE Etches
Chief Pilot: RE Tollerfield
Ceased ops end 1920 – all assets sold at auction

Avro 504K	G-EADR	Regd 6.6.19 [D6245]; sold S Summerfield & Co 3.20
DH.6	G-EAFT	Regd 1.7.19 [B2943]; CofA lapsed 22.9.20
Avro 504K	G-EAHK	Regd 18.7.19 [E4340]; sold JL Gibson 3.20
Avro 504K	G-EARZ	Regd 31.3.20; Crashed Bournemouth 7.8.20
Avro 504K	G-EASA	Regd 31.3.20; sold S Summerfield & Co 11.21
Avro 504K	G-EASB	Regd 31.3.20; CofA lapsed 1.4.21; sold CPB Ogilvie 5.25
FE.2B	G-EAHC	Regd 31.3.20; CofA lapsed 7.8.21

BRITISH EMPIRE AIR DISPLAYS
Midlands joy-riding/circus 10.4.36 – 8.36
Op by Harry Barker & Jimmy McEwan King – but fronted
[owned/financed?] by Tom Campbell-Black [see **Campbell Black
[Aviation] Ltd**]
Pilots incl: Pauline Gower [senior pilot]; Dorothy Spicer [& Senior GE]; Jock
Bonar; Robert G Doig, HGB Micklemore, Lionel Grey, Gerry Chambers,
Cressy Reynolds, Jack Wilson, Robbie Robinson & Walter Cadic
Parachutists incl Harry Ward, Bruce Williams & Bill Hire

Avro 504N	G-ACZC	Regd Earl of Cardigan
Avro 504N	G-AECR	Regd LJ Anderson; Collided with Spartan G-ABKK Westwood Heath 10.5.36
Avro 504N		Identity unknown
BAC Drone	G-AEBC	Fatal crash Hereford 1.5.36 [Walter Cadic killed]
DH.80A	G-ABYW	[pilot Blackin][identity unconfirmed]
DH.83	G-ABVI	Regd Hillman & Cross Ltd; Damaged Ashgate, nr Chesterfield 13.4.36
DH.85		Identity unknown
Hawker Tomtit	G-AEES	Regd LJ Anderson – pilot Jock Bonar
Pou du Ciel	G-ADPW	Regd/op RG Doig

Pou du Ciel	G-AEEW	Regd Aircraft Constructions Ltd/op by RG Doig
Short Scion	G-ADDT	Hired from Pobjoy Airmotors; [Gaumont British News c/s]; Crashed Porthcawl 26.7.36
Spartan 3-str	G-ABKK	Regd Air Trips Ltd/Op P Gower; Collided with G-AECR Westwood Heath 10.5.36

BRITISH FLYING & MOTOR SERVICES LTD
**The Aerodrome, Maylands Farm, Romford, Essex [based Maylands
Farm, Romford]**
Formed 22.5.28
Dirs – Mrs WE Coates [chmn], HP Cochran, DB Hope
Assets sold to AB Forsyth/RMB Ward, t/a **Inland Flying Services** 1929

DH.6	G-EBVS	Regd 23.4.28; sold Inland Flying Services 1929
DH.6	G-EBPN	Regd 23.4.28; sold Inland Flying Services 1929
Avro 548	G-EBPJ	Regd 19.10.28; Crashed Maylands 31.7.28

BRITISH HOSPITALS AIR PAGEANTS
**Promoted by Air Pageants Ltd [Jimmy McEwan King & Harry ACG
Barker]**
Tour mgr: EW Stewart; Announcer: Col J Fitzmaurice
Pilots/participants: Charles WA Scott [Chief pilot]; RH McIntosh; John BW
Pugh; Percival Phillips; Earl B Fielden; William A Rollason; Eric W
[Jock] Bonar; Leslie G Anderson; Thomas WJ Nash; Mrs Victor Bruce;
Pauline Gower; JK Morton; William AC Kingham; WHE Drury
Parachutists: 'Windy' Evans, Harry Willis; Benno de Greuw

1933 Tour – 8.4.33 – 8.10.33 [180 towns]:

Avro 504K	G-EBYW	Regd Aviation Tours Ltd; pilot Jock Bonar [identity unconfirmed]
Avro 504N	G-EBVY	Regd Leslie G Anderson [and pilot]
BAC Planette	-	Later Drone G-ADSB; painted by Heath Robinson in Chinese dragon scheme
Blackburn Lincock		Pilot – John Pugh [either G-AALH or G-ABFK of Sir Alan Cobham]
Bristol Lucifer	G-EBGA	Regd Leslie Anderson [& pilot]
Desoutter	G-AANE	Regd Walter Rickard/Rollasons; pilot Tommy Nash
DH.60X	G-EBWI	Reported as replacement by Luxury Air Tours for Fox G-ACAS
DH.60M	VH-UQA	Regd CWA Scott; pilots John Pugh/RH McIntosh
DH.60G	G-ACCY	Regd William Kingham [& pilot]
DH.82	G-ACDY	Regd Scottish Motor Traction Co Ltd; Crashed River Don, Aberdeen 26.8.33
DH.83	G-ABUP	Regd Aviation Tours Ltd; pilot Earl Fielden; Crashed Ashby, nr Scunthorpe 24.8.33
DH.83	G-ACCF	Regd CWA Scott; pilot JK Morton
DH.84	G-ACCR	Regd RT Boyd/Rollason; pilot William Rollason
Fairey Fox	G-ACAS	Regd Luxury Air Tours Ltd; pilot Mrs Victor Bruce; Forced landed & dbf Ford 14.7.33
GAL ST.4	G-ABUZ	Regd Rollason/WF Rickard; pilot TWJ Nash [possibly not used as damaged Croydon 15.4.33]
HP W.8B	G-EBBI	Aviation Tours Ltd; pilot Earl B Fielden; reportedly earmarked but not used since already wfu
Miles Satyr	G-ABVG	Regd Luxury Air Tours Ltd; pilot Mrs Victor Bruce/John Pugh; Crashed Stafford 24.5.33
Spartan 3-str	G-ACAF	Regd Hill & Phillips Ltd [Cornwall Avn Co]; pilot Percy Phillips [also possibly used G-ABWV/ABYN]
Spartan 3-str	G-ABKK	Regd Pauline Gower [& pilot]

1934 Tour "Sky Devils Air Circus"; commenced Stag Lane 15.4.34 [170
towns]
Pilots: R Robinson; EW Jock Bonar; RE Watts; Morris E Hearn; G
Williams; B Bulmore; Bogie Grey

AW Argosy	G-EBLF	Chartered by EB Fielden/Air Tours Ltd from Imperial Airways
Avro 504N	G-ACLV	Regd Air Travel Ltd; Damaged in collision with G-ACPD 30.9.34
Avro 504N		Pilot Bogie Grey
Avro 504N		
Avro Tutor		
DH.60M	G-ACOA	CWA Scott
DH.83		
Blackburn Lincock		Pilot Jock Bonar
Miles Hawk 3-str	G-ACPD	Regd TC Place; Collided with Avro 504N G-ACLV & crashed Claybury, Woodford, Essex 30.9.34

BROMPTON MOTOR CO LTD
7 Upper St Martins Lane, London WC2
Manager/pilot: Capt Arthur H Dalton
Application to operate joyriding from Southsea Common rejected by Portsmouth Council 19.4.21 & moved to IoW
Ops taken over by **Martin Aviation Co** 1922

DH.6	G-EAWT	Regd 4.21; sold Martin Avn Co 8.22
DH.6	G-EAWU	Regd 27.4.21; Crashed early 1922
DH.6	G-EAWV	Regd 4.21; CofA lapsed 11.5.22 and sold Martin Avn Co

BY AIR LTD
50 Earl St, Coventry [based Baginton]
Formed 7.19 to acquire the business of WR Johnson, TT Laker, JW Batchelor & EW Saward t/a **The Aerial Transport Co**
Pilots: Claud GM Le Champion [3.20]
Sold out to JG Riley 1920 [but who was killed in G-EALW]

BE.2E	G-EACY	Regd TT Laker 29.5.19; Crashed Lilbourne, Northants 1919
AW FK.8	G-EALW	Regd 3.9.19; Fatal crash Biddenham, Bedford 16.8.20
DH.6	G-EAQB	Regd 11.12.19; CofA lapsed 2.1.21
DH.6	G-EAQC	Regd 12.19; sold B Martin 6.20

CAMBRIDGE SCHOOL OF FLYING & AERODROME CO LTD
2 Downing St, Cambridge [based Hardwick but also joy-riding at Skegness/Hunstanton]
Proprietor: JL Lee-Jones
Pilots: Capt EE Fresson [joy-ride pilot at Skegness 7.19-9.19]; Capt R Birkbeck [joy-ride pilot at Hunstanton 19.20]; Lt FJ Ortweiler
Receiver appointed 20.4.20

Avro 504K	G-EAEC	Regd 6.6.19 [E3501]; cld/sold 8.19 – to G-AUDQ 3.22
Avro 504K	G-EAHL	Regd 18.7.19 [E4118]; sold WG Chapman 10.20
DH.6	G-EALS	Regd 29.8.19; Not certified
DH.6	G-EALT	Regd 29.8.19; CofA lapsed 19.12.20

CD BARNARD AIR TOURS LTD
Room 625, Grand Buildings, Trafalgar Square, London WC2
Formed 3.31 by Charles D Barnard & Mrs ME Barnard
Circus tour of England, commencing 1.4.31 [with 4 acft] for 118 towns [150 intended]; ending 11.10.31
Pilots: Charles Barnard, Edward D Ayre [dep 8.31], FS Crossley; Leonard H Stace [joined mid 4.31]; T Neville Stack [joined 8.31]; C Reilly
Many of Circus to Skywork Ltd 12.31 for South Africa tour
Reformed for tour to India as India Air Pageants Ltd; departed 11.11.33

Fokker F.VIIA	G-EBTS	Operated 4.31; Regd CD Barnard 2.32 *"The Spider"*; sold India Air Pageants Ltd 9.34
Spartan 3-str	G-ABJS	Hired from Spartan Aircraft Ltd 6.31; pilots TN Stack/C Riley
Spartan 3-str	G-ABET	Unconfirmed
Cierva Autogiro	G-AAYP	Hired from Cierva Autogiro Ltd 6.31; pilot Reginald Brie
Cierva C.19	G-ABFZ	Hired from Cierva Autogiro Ltd 8.31; pilot Reginald Brie
DH.60X		Hired from National Flying Services Ltd [but possibly DH.60G with Jim Mollison?]
Desoutter		Hired from National Flying Services Ltd; pilot Joe King
Sports Avian	G-AAXH	Regd Henlys Ltd *"The Red Herring"*; pilot Len Stace
Potez 36	F-ALJC	Regd [as G-ABNB] to Lord Halsbury; pilot Bud Cummings

THE CENTRAL AIRCRAFT CO
Palmerston Works, 179 High Road, Kilburn, London NW6 [based Kingsbury 19.20; to Northolt]; joy-riding over London, Kent and The Solent 1920
Formed 1916 – subsidiary of R Cattle Ltd, woodworkers
Pilots: Herbert Sykes [chf pilot – joy-riding]; Leslie Castlemain; Capt Anderson; HH Maddox; EB Wilson

Avro 504K	G-EADO	Regd 5.6.19 [F8706]; Crashed Clayton West 10.19
Avro 504K	G-EADP	Regd 5.6.19 [E3481]; *"White Wings"*; sold Midland Avn Co 10.21
Avro 504K	G-EADQ	Regd 5.6.19 [D7588]; sold to IFS Air Corps as *"VI"*

Avro 504K	G-EADU	Regd 6.6.19 [D9329]; sold JM Drysdale 9.20
Avro 504K	G-EAFP	Regd 25.6.19 [E3363]; cld/sold 12.21 – to G-AUFP 12.23
Avro 504K	G-EAGI	Regd 9.7.19 [E4234]; Crashed Marble Hill, Twickenham 19.3.20
Avro 504K	G-EAGJ	Regd 9.7.19 [C746]; Crashed Twickenham 12.19
Avro 504K	G-EAJK	Regd 8.19; sold Vulcan Avn Co Ltd 5.21

CHELMSFORD & DISTRICT AERO CLUB
Broomfield Aerodrome, Chelmsford, Essex
Opened 18.5.32
Proprietors: Capt Heinz M Talbot-Lehmann, George W Higgs [resigned 9.32] of **Aviation Transport Sales & Service Ltd**
Also operated circus/barnstorming wef 6.32 in locality 1-2 days/week – Avro 504K [HMT-L]; Avian/Klemm [LG Rumsey]; Klemm #2 [John Rogers] & AW Fairlie [para]; Jack Harris [wing-walker]; Tommy Bateman [motorcyclist]

Avro 504K	G-ABYB	Regd HM Talbot-Lehmann; CofA lapsed 28.2.33
Avian [Hermes II]		Op 6.32; possibly G-AATL [regd GW Higgs; regd ATS&S Ltd 8.32]
Klemm [Salmson]		Op 6.32; identity not known
Klemm		Op 6.32; Identity not known

COBHAM & HOLMES AVIATION CO
See **Berkshire Aviation Co**

THE CORNWALL AVIATION CO [LTD]
Truro Road. St Austell, Cornwall [based Rocky Parc, St Austell]
Formed 5.24 as subsidiary of **Hill & Phillips Ltd** [garage owners, St Austell] by FL [Leonard] Hill & Capt Percival Phillips
Albert J Adams [GE/Mgr; dep to Aeroplane Services Ltd 3.29; returned but left again 1931]
Company formed 3.9.24
Pilots: Capt EW Jordan [3.26; dep 3.27]; Sammy Summerfield [1927-29]; Don P 'Jock' Cameron [1927-29]; Albert J Adams [GE 1924; pilot 9.27]; C Bint [1928]; Ernie Bicknell [1930]; Capt H Lawson [appt 1930]; Capt E Stewart [appt 1931]; JC Higgins [1931]; Capt ED Crundall [appt 1931]; A West [1931]; Martin Hearn [1932]; Capt JD Parkinson [1932]; WHP Stewart [1932]; Capt Glover [1932]
Fire burnt down tent/hangar at St Austell, destroying 'several' Avro fuselages
Also used Sopwith Snipe H354 fuselage as engine test-bed behind garage
Receiver appointed 11.32

Avro 504K	G-EBIZ	Regd Hill & Phillips Ltd 29.4.24; CofA lapsed 23.3.33
Avro 504K	G-EBNR	Regd 1.3.26; CofA lapsed 13.2.30
Avro 504K	G-EBSE	Regd 4.7.27; CofA lapsed 11.4.32
Avro 504K	G-AAAF	Regd 8.28; Badly damaged Abergavenny 6.32, sold in crashed state 11.32; to Surrey Flying Services Ltd 8.33
Avro 504K	G-AAYI	Regd 5.5.30; CofA lapsed 28.5.31
Parnall Elf	G-AAFH	Loaned from Geo Parnall & Co 1931
Avro 504K	G-ABHI	Bought 4.32; CofA lapsed 24.5.33
Avro 504K	G-AAUJ	Regd P Phillips 24.6.32; Fatal crash Harrogate 5.10.32
Spartan 3-str	G-ACAF	Regd Hill & Phillips Ltd 3.33; sold Air Pageants Ltd 9.36

CORONATION AIR DISPLAYS
Trading name for **Aircraft Demonstrators Ltd** 4.37-8.37

CRIMSON FLEET CIRCUS
Trading name for **Modern Airways Ltd**

CWA SCOTT'S FLYING DISPLAY LTD
Grand Buildings, Trafalgar Square, London WC2 [based Croydon/Hanworth]
Formed 20.12.35 by Charles WA Scott & Capt Percival Phillips [ex Cornwall Avn Co]; to take over assets/acft/goodwill from **National Avn Displays Ltd** [wef 11.12.35]
DL Eskell [appt Genl Mgr]; Announcer: Roy Arthur
Pilots: J King [Ferry]; Percival Phillips; Launcelot J Rimmer [Avro Mongoose/Scion]; Idwal Jones [Drone]; Mrs Winifred Crossley [G-ADWG]; HA Shotter [Cadet]; AL Harris; RJ Ashley [C.30]; Martin Hearn
Receiver appointed 23.9.36

AS Ferry	G-ABSI	dd 31.12.35; Regd 12.3.36; sold Air Publicity Ltd 11.36
AS Ferry	G-ACFB	Regd 7.4.36; sold Air Publicity Ltd 11.36

Avro 504N	G-ACOK	Regd P Phillips
Avro 504N	G-ACRE	Regd P Phillips
Avro 504N	G-ADDA	Regd P Phillips
Avro 640 Cadet	G-ACFU	Regd Tom Campbell-Black; operated 4.36
Avro 640 Cadet	G-ACLU	Regd 12.3.36; sold A Harris 12.36
Avro 640 Cadet	G-ACOZ	Regd 12.3.36; CofA lapsed 5.4.37
Avro 640 Cadet	G-ACPB	Regd 12.3.36; sold JL Bebb 2.37
BAC Drone	G-AEEO	Regd 1.4.36; sold LJ Rimmer 10.36
BAC Drone	G-AEAN	Regd BAC [1935] Ltd; Crashed nr Southend Airport 22.7.36
BA Swallow		Not identified; operated 4.36
Cierva C.30A	G-ACUT	Regd 5.3.36; sold Malcolm & Farquharson Ltd 3.37
DH.82	G-ADWG	Regd 3.3.36; sold Cinque Ports F/C Ltd 2.37
Pou du Ciel	G-AEFK	Regd 30.3.36 *"Bertie Bassett"*; CofA lapsed 30.9.36
Pou du Ciel	G-ADSC	Regd Scott Motors [Saltaire] Ltd; operated 4.36
Praga Baby	G-ADXL	Regd F Hills & Sons Ltd; operated 4.36 [pilot A Harris]
Short Scion	G-ACJI	Operated 1936 [by Lance Rimmer]
Southern Martlet	G-ABIF	Regd Air Travel Ltd – unconfirmed
Wolf Glider		Operated 4.36 – Mrs Ronald Price [Miss Joan Meakin]

DEVONSHIRE AVIATION TOURS
7 Thompson Rd, Exeter [based Exeter]
Prop: Hayward

Avro 504K	G-ABZC	Regd 7.4.33; Fatal crash Chard, Somerset 30.4.33
Avro 504K	G-AAYM	Regd 7.4.33; sold A Tapp 7.33

DOMINION AIRCRAFT LTD
Hand & Sceptre Hotel, Southborough, Tunbridge Wells [based Gatwick?]; later 3 Liverpool Gardens, Worthing [based Shoreham]
Operation commenced summer 1928 by John C Don & FS Miles
Coy formed 26.11.28
Operated in South Africa 1928/29

Avro 504K	G-EAJU	Regd JC Don 7.8.28; Crashed Brighton 21.7.29
Avro 504K	G-AACW	Regd 19.11.28; No CofA issued; sold Southern Acft Ltd 6.29
Avro 504K	G-AACX	Regd 11.28; regn cld 4.29

THE EASTBOURNE AVIATION CO LTD
The Crumbles, Eastbourne [based Eastbourne]
Formed 1911; company formed 18.2.13
Fndr/dir: Maj Frederick B Fowler [MD]
Joyflights Eastbourne/Bexhill/Worthing [using Short 184/Avros]
Flying ceased Spring 1921 & receiver appointed 16.12.22

Avro 504L	G-EAFB	[E.1]; Regd 18.6.19 [K-144]; CofA lapsed 24.6.21
Avro 504K	G-EAJG	Regd 7.8.19; CofA lapsed 9.7.21; sold T Baden Powell 3.22
Avro 504L	G-EAJH	[E.2]; Regd 7.8.19; [based Brighton Beach]; Crashed & sank off Hove 19.8.20
Short 184	G-EAJT	Regd 28.8.19 [N2986]; regn cld 8.20
Short 184	G-EALC	Regd 28.8.19 [N2998]; CofA lapsed 6.3.20
Avro 504K	G-EALD	Regd 8.19 [H1925]; CofA lapsed 9.6.21; sold RH Leavey 5.22
Avro 504L	G-EALO	[E.3]; Regd 26.8.19; CofA lapsed 4.9.20
Avro 504L	G-EANS	[E.4]; Regd 26.9.19; regn cld 9.20
Avro 504L	G-EASD	[E.5]; Regd 26.3.20; sold to Sweden as S-AAP 8.21, later S-AHAA
Avro 504L	G-EASE	[E.6]; Regd 26.3.20; regn cld 2.3.21 – to Sweden?

ESSEX AVIATION CO
Thornwood, Epping, Essex [based Epping]
Prop/pilot – Frank T Neale
Joyriding at Cliftonville 8.23-9.23

Avro 504K	G-EBCK	Regd FT Neale 1.23; sold B Roberts 2.25

FLYING FAIR
See **Aviation Developments Ltd**

G & H AVIATIONS LTD
Meadow Works, Pricklers Hill, Great North Rd, Barnet [based Stag Lane/Barnet]
Formed 5.30 as **G&H Aviation** and Company formed 8.9.30
Fndr/dirs: F Greenwood, JW Halliwell, H Whitehead [all former ventilating engineers]
Avro 504Ks were built up from spares at Stag Lane 1930

Avro 504K	G-ABAA	Regd 5.30; sold Luffs Aviation Tours 3.33
Avro 504K	G-ABAB	Regd 5.30; sold RO Roch 6.31

THE GIRO AVIATION CO
337 Royal Liver Buildings, Liverpool [based Hesketh Park Aerodrome, Southport]
Prop/pilot: Sidney Norman Giroux

DH.6	G-EARC	Regd 25.5.20; CofA lapsed 3.8.24
DH.6	G-EAVG	Regd 16.8.20; Forced landed in mudflats in Ribble Estuary and set on fire 1921; salvaged but not rebuilt
DH.6	G-EARK	Regd 24.3.21; CofA lapsed 12.5.22
DH.6	G-EARM	Regd 24.3.21; CofA lapsed 19.4.23
DH.6	G-EARJ	Regd 12.7.21; CofA lapsed 3.7.23
DH.6	G-EBEB	Regd 11.7.22; *"Maybus"*; CofA lapsed 13.9.30; [burnt Hesketh Pk in WWII]
DH.6	G-EBWG	Regd 21.6.29; *"Silver Wings"*; CofA lapsed 15.4.31; [burnt Hesketh Pk in WWII]
Avro 548	G-EAFH	Regd 27.3.31; Crashed Southport 31.5.35

Avro 548	G-ABMB	Regd 7.5.31; CofA lapsed 17.5.35
Avro 548	G-ABSV	Regd 22.12.31; CofA lapsed 31.8.35
Avian IIIA	G-EBZM	Regd 9.12.32; CofA lapsed 20.1.38; stored Hesketh Park, Southport; to Bernard Murphy, Ringway for rebuild early 1950s
DH.83	G-ACCB	Regd 1.1.36; Stored Hesketh during WWII and CofA renewed 2.4.47
DH.83	G-ACEJ	Regd 22.7.36; Stored during WWII and CofA renewed 5.6.46

GOLDEN EAGLE AVIATION CO LTD
438 Corn Exchange Buildings, Corporation St, Manchester; later to 3 York St, Manchester [based Marton Moss, St Annes-on-Sea/Squires Gate]
Formed 8.19 by Maj Horace S Shield [MD]; Lt A Pearson & Lt WC Ellis
Pilot: Capt Hedges

Avro 504K	G-EAFX	Regd 1.7.19 [H6543]; W/off in gales 29.8.19 prior to overhaul and rebuilt as G-EASG
DH.6	G-EAHD	Regd 17.7.19 [B2934]; regn cld 2.20
DH.6	G-EAHE	Regd 18.7.19 [B2917]; Crashed 25.2.20
DH.6	G-EAPG	Regd 8.11.19; regn cld 11.20
DH.6	G-EAPH	Regd 8.11.19; No CofA issued; regn cld 2.20
DH.6	G-EARJ	Regd 3.20; sold Giro Avn Co 7.21
DH.6	G-EARK	Regd 3.20; sold Giro Avn Co 3.21
DH.6	G-EARL	Regd 3.20; sold HB Elwell 12.20
DH.6	G-EARM	Regd 3.20; sold Giro Avn Co 3.21
Avro 504K	G-EASG	Regd 3.20; No CofA issued; sold JS Boumphrey 7.21

THE GRAHAME-WHITE CO LTD/
GRAHAME-WHITE AVIATION CO LTD
The London Aerodrome, Hendon NW9 [based Hendon]
Formed 11.09 by Claude Grahame-White as aircraft manufacturer and also flying training.

Operated joy-riding in early 1920s – chief pilot H Chamberlayne
Liquidator appt 1926.

Avro 504K	G-EAAX	Regd 6.5.19 [D6205]; CofA lapsed 11.5.21
Avro 504K	G-EAAY	Regd 6.5.19 [C724]; CofA lapsed 25.8.22; sold Lady U Beecham 9.24
Avro 504K	G-EABA	Regd 6.5.19 [E3480]; CofA lapsed 28.5.21
Avro 504K	G-EABE	Regd 6.5.19 [E4137]; CofA lapsed 13.7.22; sold EF Edwards 9.23
Avro 504K	G-EABF	Regd 6.5.19 [C748]; CofA lapsed 8.7.21
Avro 504K	G-EABG	Regd 6.5.19 [C749]; CofA lapsed 31.5.21
Avro 504K	G-EABH	Regd 6.5.19 [B8758]; CofA lapsed 1.8.21
Avro 504K	G-EABN	Regd 10.5.19 [C723]; CofA lapsed 8.5.21
Avro 504K	G-EABX	Regd 10.5.19 [E4230]; sold CB Wilson 8.22
Avro 504K	G-EABP	Regd 19.5.19 [F9802]; sold to Belgian AF 4.21
Avro 504K	G-EABO	Regd 20.5.19 [D6202]; sold to G CAAE 7.20
Avro 504K	G-EABW	Regd 20.5.19 [C747]; CofA lapsed 18.7.22
Blackburn Kangaroo	G-EADE	Regd 8.6.19 [B9981]; Crashed Hendon 29.6.19
Blackburn Kangaroo	G-EADF	Regd 8.6.19 [B9982]; Crashed Hendon 31.5.19
Blackburn Kangaroo	G-EADG	Regd 11.6.19 [B9985]; CofA lapsed 7.6.21
DH.6	G-EAGE	Regd 4.7.19 [C5224]; CofA lapsed 5.9.21; sold CB Wilson 8.22
DH.6	G-EAGF	Regd 4.7.19 [C5220]; CofA lapsed 21.8.21; sold The Hon E Mackay 10.21

INDIA AIR PAGEANTS LTD
170/173 Piccadilly, London W1; Bank of Baroda Building, Apollo St, Bombay [based Bombay]

Formed out of **CD Barnard Air Tours Ltd** by CD Barnard
Circus dep for India 11.11.33; arr Bombay 8.12.33; initial show Poona 18.12.33 or Bombay 23.12.33; tour completed 4.34 [92 displays]
Capt AH Dalton [Genl Mgr]
G-EBTS flown out; arr Bombay 15.12.33; 7 others shipped out, dep Heston 2.12.33
Messrs Wyndham, Joyce and Ward were parachutists

Fokker F.VIIA	G-EBTS	Regd CD Barnard Air Tours Ltd; regd to company 25.9.34 *"The Spider"*; sold Sir DH Bhiwandiwalla 5.35
BAC Drone		Shipped out late 11.33; [flown by JC Longmore]; Crashed Dum Dum 1.34
DH.82	G-ACFA	Regd CD Barnard; Op Jock Mackay; sold as VT-AGD 1.35
DH.83	G-ACKZ	Op RL Palmer
DH.83		Op JR Hatchett
DH.83		Op JB Pugh
DH.83		Op A Auping; dep late 11.33
Miles Satyr	G-ABVG	Op JBW Pugh
Segrave Meteor		Op WA Burnside
Spartan 3-str		Op WR Andrews

INGHAM & LITTLE AVIATION CO
Lynwood, Penrith [based Carlisle/Oliver Mount, Scarborough; to Cockermouth/Heysham [1921]]

Joyriding coy; successor to **Border Aviation Co Ltd** 10.20 formed by RF [Bob] Little & Graham Little [brothers]; Percy Hall Ingham
Pilots: Major Ferrand [8.21-10.21]; Sidney F Woods [appt 9.21]

Avro 504K	G-EANQ	Regd 25.5.21; Crashed Carlisle/Solway Firth 1.12.21
Avro 504K	G-EAIA	Regd 6.21; CofA lapsed 15.8.21; sold R Atkinson 3.23

INLAND FLYING SERVICES [LTD]
Maylands Farm Aerodrome, Romford, Essex; to Apse Manor Aerodrome, Shanklin, IoW 8/29

Formed 1928 by Archer 'Art' Forsyth & Randall MB Ward
Company formed 6.4.29 after proprietors bought British Flying & Motor Services Ltd

Name changed 16.4.30 to **WIGHT AVIATION LTD**

Avro 548	G-EBPJ	Regd 5.28; Crashed Maylands 31.7.28 and sold to British Flying & Motor Svs Ltd 10.28
Avro 504K	G-AAFE	Regd 6.5.29; Wfu early 1930
Avro 504K	G-AAFT	Regd 6.5.29; regn cld 3.31
DH.6	G-EBPN	bt 1929 ex BMFS but not regd
DH.6	G-EBVS	bt 1929 ex BMFS but not regd

INTERNATIONAL AVIATION CO
Castle Mona, Douglas, IoM

Pilots: LJ Rimmer; S N Giroux [1919/20]

DH.6	G-EARA	Regd 11.2.20; Crashed off Douglas, IoM 25.5.20
DH.6	G-EARB	Regd 11.2.20; Damaged in gales 1920 – CofA lapsed 30.3.21
DH.6	G-EARC	Regd 11.2.20; sold Giro Avn Co 5.20
DH.6	G-EARD	Regd 11.2.20; Crashed in Irish Sea off Ramsey, IoM 3.8.20

IRISH AIR LINES LTD
14 Barronstrand St, Waterford [based Waterford]

Formed 3.32 – joy-riding business in East Cork/Waterford area by Arthur Westcott Pitt, James M StJohn-Kearney; EG Stewart; Company formed 13.6.32
Pilots: Eric G Stewart [appt chf pilot 4.32]; Andy Woods [7.32]
Liquidator appt [prior to 6.33]

Avro 504K	EI-AAM	Regd 15.3.32; Crashed Tramore Strand, Waterford 19.8.32
Avro 504K	EI-AAN	Regd 23.3.32; wfu 1932
Bluebird IV	EI-AAO	Regd 20.5.32; Crashed Tramore Strand, Waterford 31.7.32

JAMES BUNNING LTD
St Brannocks, Glasllwch, Newport, Mon; [based Prestatyn]; later National Provincial Chambers, Pontypool, Mon 9.29 [based Newport, Mon]

Formed 3.28 by RJ Bunning/Mrs Hilda F Bunning as joy-riding business

Avro 504K	G-EBSM	Regd 15.5.28; CofA lapsed 17.5.29

JUBILEE AIR DISPLAYS LTD
Salisbury Square House, Fleet St, London EC4

Name changed from **Aviation Displays Ltd** 5.2.35 on reformation of **Air Pageants Ltd**
Dirs: Leslie G Anderson; James McEwen King; Henry ACG Barker
Chief pilot Lt Owen Cathcart-Jones; others included ME Hearne, J King, HG Bulmore, RF Robinson, EW Bonar, Carleton Ross & 'Stainless' Steele. Parachutists were Harry Ward & Bill Hire
Operated UK/Ireland Tour 4.35-9.35
Probably also operated **British Empire Air Displays** 4.36-8.36

Avro 504N	G-ADBS	Regd LG Anderson; Fatal crash Bodmin 16.8.35
Spartan 3-str	G-ABKK	Regd Air Trips Ltd
Monospar	G-ADDZ	Pilot Cathcart-Jones
Short Scion	G-ADDT	

KINGSBURY AVIATION CO
Church Lane, Kingsbury Green, London

Formed 10.5.17 – subsidiary of Barningham Ltd & built DH.6/Sopwith Snipes 1918/19
Fndr/dirs: Ernest Barningham; Warwick Wright [motor dealer]; Charles Lane [of Charles Lane & Co, aircraft parts manufacturer]
Pilot: Capt James Cordes [pilot 1919]
Name changed to **KINGSBURY ENGINEERING CO LTD** 1919 and diversified into motor scooters & cars
Liquidator appointed 1921

AW FK.3	G-EAEU	Regd 18.6.19 [B9612]; regn cld 12.19
Avro 504K	G-EAGT	Regd 14.7.19 [D6217]; sold Astra Eng'g Works & Garage 5.20
DH.6	G-EAKU	Regd 8.19; sold Astra Eng'g Works & Garage 5.20

KINGWILL & JONES FLYING CO
Alloa Aerodrome, Alloa; later Turnhouse Aerodrome, Crammond Bridge, Midlothian

Formed 1921 by Edward A Jones [to Brooklands School of Flying 11.28] & Archibald N Kingwill [to Wm Beardmore 1923]
Operated Withernsea/Yorkshire coast
Pilot: Capt Geoffrey R Beck

Avro 504K G-EADH with co-owner Edward Jones.

| Avro 504K | G-EBCB | Regd 16.3.22; Destroyed in gales 11 28 |
| Avro 504K | G-EADH | Regd 13.5.22; CofA lapsed 17.6.26 |

KINMEL BAY FLYING SERVICES
13 The Ridgeway, Enfield, Middlesex [based Kinmel Bay, Rhyl]
Prop: Maj Edward G Clerk
Operations during 1934 prior to owner taking circus to Australia

| Avro 504K | G-EBYW | Regd EG Clerk 21.6.34; to Australia, crashed Stanthorpe, Qld 30.1.35 |

THE LAKES MOTOR & SEAPLANE CO
Cockshott, Lake Windermere
Formed 1920 by Howard Pixton to take over Avro facility previously run by him
Possibly operated Avro 504Ls G-EADJ & EADK

LEATHERHEAD AVIATION SERVICES
Leatherhead Motor Works, Kingston Rd, Leatherhead [based Croydon/Byhurst Farm, Malden Rushett, Leatherhead]
Prop: WG Chapman
Pilots: Capt AF Muir [to Surrey F/Svs 2.21]; FH Haynes [4.21; 9.21]; Knowles [4.21]
Ceased trading 8.22 after Chapman injured in crash of G-EBAV

DH.6	G-EANU	Regd Leatherhead Motor Co 9.19; sold JV Yates 5.2
Avro 504K	G-EAHL	Regd WG Chapman 9.20; sold Manchester Aviation Co Ltd 3.23
Avro 504K	G-EBAV	Regd 2.1.22; Crashed Slough 20.8.22
Avro 504K	G-EBCQ	Regd 4.22; Crashed Surbiton 7.7.22; sold Manchester Avn Co Ltd 12.22 and rebuilt

LINCOLNSHIRE FLYING SERVICES
158 Eastgate, Louth, Lincs; Kenwick Road, Louth [8.39] [based Kenwick Road, Louth]
Formed 8.34 by FG Wright & Craig

| Avro 504N | G-ADFW | Regd 4.4.35; CofA lapsed 13.5.37 |

THE LIVERPOOL AVIATION CO
68 Victoria St, Liverpool; later 53 Derby Lane, Stoneycroft, Liverpool

| Avro 504K | G-EAAY | Regd 12.5.25; sold Southern Counties Avn Co 4.26 |

LJ SKY TRIPS LTD
41 Baronsmere Road, E Finchley N12 [based Shoreham; Bekesbourne 7.28] based Folkestone 7.29?
Formed 16.7.27 by Leslie A Lewis & LA Jackson and undertook itinerant joy-riding in Home Counties

Avro 504K	G-EAAY	Regd 20.7.27; sold LR Gladwin Errington 7.30
Avro 504K	G-EBWF	Regd 2.28; sold Alliston Avn Co Ltd 6.29
Avro 504K	G-EBZA	Regd 7.28; sold DW Sch of Flying Ltd 7.30

LLOYDS COMMERCIAL AIRCRAFT CO
124 Baker St, London W1 [based Brooklands/East Anglia]
Prop: George M Lloyd

| Sopwith Gnu | G-EADB | Regd 19.2.26; Crashed Horley 2.3.26 |
| Sopwith Gnu | G-EAGP | bt.1926 [not regd]; Fatal crash Kings Lynn 2.5.26 |

LYTHAM AVIATION CO
Lytham, Lancs [possibly operated from Sands Cottage Estate, Bridlington]

| Avro 504K | G-EASG | Regd 7.22; sold Southern Counties Avn Co 5.25 |

MANCHESTER AVIATION CO
41 John Dalton St, Manchester; to Royal London Buildings, 196 Deansgate, Manchester [based Alexandra Park] [based Chorlton-cum-Hardy Aerodrome 4.23]
Formed 1920 by Charles Bernard Wilson [also involved Roland Turner?]
Pilots: Newnham [4.23]; VN Dickinson [appt chf pilot 4.23]

DH.6	G-EAHI	Regd 29.10.20; CofA lapsed 30.7.22
Avro 504K	G-EAZW	Regd 12.21; sold Northern Avn Co 3.24
Avro 504K	G-EAZX	Regd 12.21; sold Northern Avn Co 3.24
DH.9C	G-EBDG	Regd 5.22; sold Northern Avn Co 3.24
Avro 504K	G-EABX	Regd 2.8.22 CB Wilson; CofA lapsed 21.5.22
Avro 504K	G-EBCQ	Regd 18.12.22; Crashed Barrow-in Furness 24.8.23
Avro 504K	G-EAGC	Regd 8.3.23; CofA lapsed 9.11.22 and not used
Avro 504K	G-EAHL	Regd 8.3.23; Crashed Dunham-Massey, Cheshire 30.6.23
Avro 504K	G-EADP	Regd 8.3.23; sold Northern Avn Co 3.24
Avro 504K	G-EBGH	Regd 8.5.23; Crashed Broughton Grange, Brigg 5.6.23
DH.6	G-EAUS	bt.1923 for spares [not regd]

THE MARTIN AVIATION CO
The Cliff Garage, Cleethorpes; also Addison Lodge, Gedling, Notts [based Gedling 6.24]
Prop: Capt Bernard Martin [also possibly A Martin & AD Martin – no relation?]

Lincolnshire Flying Services Avro 504N in a typical rural scene.

One of the DH.6s operated by The Martin Aviation Co.

Joyriding from East Coast/Cleethorpes; took over ops of **Brompton Motor Co Ltd** in IoW 8.22

DH.6	G-EAQC	Regd B Martin 3.6.20; regn cld 6.11.21
Avro 504K	G-EAOE	Regd B Martin 30.5.22; Crashed Cleethorpes Foreshore 12.6.22
DH.6	G-EAWT	Regd 4.8.22; CofA lapsed 27.8.24
DH.6	G-EAWU	Not regd; crashed 1922
DH.6	G-EAWV	Regd 10.6.24; No CofA issued

MIDLAND AERO FLIGHTS LTD
4 Blenheim Road, Moseley, Birmingham; 31 Temple Row, Birmingham [1929] [based Studley]
Formed 4.7.28
Dirs: H Oldfield [MD], HD Hands [dir/stockbroker]

Avro 504K	G-EBYE	Regd 7.28; sold RH Jackson 11.28

MIDLAND & SCOTTISH AIR FERRIES [LTD]
Moorpark Aerodrome, Renfrew [opened base at Hooton 6.33] - airline, air charter and joyriding from various locations, including Blackpool/Stanley Park and the beach at Ayr
Formed 12.32 by John Sword - manager of Western division of Scottish Motor Traction, but as a separate private venture; company formed 10.3.33
Dirs: John C Sword [MD]; Mrs Christina G Sword; Manager - J Graham McDonald; Secretary - Miss Mary Drinkwater
Pilots: Sqn Ldr Harold GR Malet [chief pilot]; John P Rae [to chief pilot/operations manager 6.33]; JH 'Jimmy' Orrell; SW 'Ned' Sparkes; MN Mavrogordato; EG 'Ed' Stewart; Cyril HA Colman; David Barclay; Hatchett; George Walker; Miss Winifred <u>Joyce</u> Drinkwater [appointed pilot 6.33; sister of above]; Flt Lt APK Hattersley [appointed pilot 2.34]; F/O Charles N Pelly [appointed pilot 2.34]; RM Saundby [appointed pilot 2.34]
Operations ceased 30.9.34 after the 'railway' interests on SMT objected to competition on newly-formed Railway Air Services

Airspeed Ferry	G-ACBT	Regd 21.12.32; to long-term storage at Renfrew 12.3.35
DH.83	G-ACBZ	Regd 24.1.33; CofA lapsed 9.3.35 and sold to VH-UZD 4/37
DH.83	G-ACCB	Regd 24.1.33; [based Stanley Park wef 21.5.34]; sold Giro Aviation Co 1.36
DH.83	G-ACCT	Regd 13.2.33; [based Stanley Park wef 14.5.34]; sold West of Scotland Air Services 7.35
DH.83	G-ACCU	Regd 13.2.33; CofA lapsed 4.4.35 and sold to VH-UZC 3.37
Airspeed Ferry	G-ACFB	Regd 6.3.33; CofA lapsed 5.4.35 and sold CWA Scotts Flying Displays Ltd 4.36
Spartan Arrow	G-AAWZ	Regd 10.4.33; sold Brian Lewis & Co Ltd 2.35
Avro Ten	G-ACGF	Regd 11.4.33; Withdrawn from use 25.7.34 and to long-term storage at Renfrew
Avro 640 Cadet	G-ACFX	Regd 12.4.33; CofA lapsed 15.1.35 and sold to VR-RAJ 1.36
Avro 642 Eighteen	G-ACFV	Regd 12.4.33 *"Marchioness of Londonderry"*; damaged 4.6.34 and sold to Commercial Air Hire Ltd 5.35
DH.84	G-ACCZ	Regd 21.4.33; sold Crilly Airways Ltd 5.35
DH.84	G-ACDL	Regd 21.4.33; sold Mrs V Bruce/ Commercial Air Hire Ltd 9.34
DH.84	G-ACDM	Loaned ex SMT 5.33-11.33
Avro 643 Cadet	G-ACIH	Regd 7.33; sold Marquess of Londonderry 6.34
DH.84	G-ACJS	Regd 2.8.33 [Operated as J Sword's personal aircraft]; sold Northern & Scottish Airways Ltd 1.35
DH.84	G-ACDN	Dd ex SMT 22.9.33; sold Crilly Airways Ltd 5.35
DH.84	G-ACET	Dd ex SMT 15.2.34; sold Highland Airways Ltd 9.34
DH.60GIII	G-ACGD	Bt 4/34 [based Stanley Pk wef 6.34]; sold Blackpool & Fylde A/C 12.34

MIDLAND AVIATION CO
17 Craven St, Melton Mowbray [based Castle Bromwich; also joyriding at The Denes, Great Yarmouth]
Prop: HW Barrett
Pilots: RH Leavey [6.23]; Capt Geoffrey R Beck

DH.6	G-EARR	Operated 1921; CofA lapsed 27.5.21

Avro 504K	G-EADP	Regd 13.10.21; sold Manchester Avn Co 3.23
Avro 504K	G-EAGB	Regd 19.8.22; Crashed Evesham 10.9.23
Avro 504K	G-EALE	Regd 19.8.22; Crashed Burton, Cheshire 30.3.23

THE MIDLAND AVIATION SERVICE
Sunny Vale, Rockley, Retford, Notts [based Retford]
Prop: C Hudson
Pilot: Capt Jackson

DH.6	G-EARR	Regd C Hudson 21.5.21; CofA lapsed 27.5.21

One of the crimson-painted Modern Airways Avro 504Ns.

MODERN AIRWAYS LTD
27 Beaufort Gardens, London SW3 [based Knotts Farm, Kingsdown, Wrotham, Kent 3.32; also Porthcawl]

Formed 20.2.32 by Mervyn AP Rhys Pryce [MD], S Caplowe, JI Pickering and traded as **The Crimson Fleet**

Opening display 5/6.5.32 at Buckhurst Park, Withyham, Kent

Pilots: Miss Pauline Gower/Dorothy Spicer; Cyril R Cubitt [5.32]; EHM Slade [5.32]

Avro 504N	G-EBHT	Regd 19.3.32; Crashed Locks Common, Porthcawl 25.6.32
Avro 504N	G-EBHD	Regd 19.3.32; CofA lapsed 2.3.33
DH.60G		Op 5.32 [prob G-AAKG]
Spartan 3-str	G-ABKK	Regd Pauline Gower; Damaged in collision DH.60G G-AAKG Cardiff 8.6.32
DH.60G	G-ABFD	Probably hired in 6.32 after accident to G-ABKK [also suggested as Spartan G-AAGY]
Avro 504K	G-ABSL	Regd 8.32; sold GW Cannon/FA Manning 5.33

NATIONAL AVIATION DAY LTD
Grand Buildings, Trafalgar Square, London WC2 [based Stag Lane, Croydon; to Ford 10.32]

Formed 22.1.32 by Sir Alan J Cobham; to op National Avn Day displays in 175 towns

Dallas Eskell [Appt Genl mgr]

1932 Tour launched at luncheon at Connaught Rooms 4.4.32; tour commenced Luton 13.4.32; ended 16.10.32 [170 towns actually visited]

Pilots: Hugh C Johnson [chf pilot/HP W.10], EB Fielden, Flt Lt CK Turner-Hughes, TWJ Nash [Desoutter], AF Muir, W Bonar, P Phillips, H Lawson, Charles WH Bebb. Joe R King, R Ogden, JD Parkinson, Capt EW Stewart; ED Crundall, JJ Scholes, Sqn Ldr Mallet; AC Rawson; Martin Hearn [wing-walking], F Kent; also I Price & Naomi Heron-Maxwell as parachutists

To **South Africa** 10.32 for Winter tour. [Tour mgr Dallas Eskell: Participants: Alan Cobham, Lady Cobham, Flt Lt CK Turner Hughes; Flt Lt AHC Rawson, HC Johnson [chf pilot], Harold Lawson, Charles WH Bebb, Martin Hearn; R Ogden with I Price as parachutist. Arr Cape Town 1.12.32

1933 Tour 14.4.33-7.10.33; two similar 12 aircraft teams, each covering 165 towns [actual 306 total]. Pilots incl Charles K Turner-Hughes, AHC Rawson; AN Kingwill; Charles WH Bebb; William 'Jock' Mackay; HC Johnson [chief pilot], JR King, G Tyson, B Wilson, F Jacques, R Ogden, CH Bremridge, R Warner, W Easdown, W 'Dusty' Miller, George Kemp, LJ Rimmer, GV Tyson, P Phillips, JD Parkinson; also I Price & Harry Ward as parachutists and Martin Hearn as wing-walker

1932 Tour – Aircraft regd to Company or other Alan Cobham entities:

Airspeed Ferry	G-ABSI	Regd Sir Alan Cobham 12.31 *"Youth of Britain II"*; regd National Aviation Displays Ltd 4.4.35; sold CWA Scotts Flying Display Ltd 3.36
Airspeed Ferry	G-ABSJ	Regd Sir Alan Cobham 12.31 *"Youth of Britain III"*; later *"Youth of Africa'*; regd National Aviation Displays Ltd 27.11.33; sold VT-AFO 4/34
Cierva C.19/IVP	G-ABGB	Regd Sir Alan Cobham 11.4.32 [pilot AC Rawson]; Crashed Cape Town 13.2.33
DH.9	G-AACR	Regd 21.6.32; CofA lapsed 21.6.33
DH.60G	G-ABUB	bt 29.7.32 by Sir Alan Cobham; regd to him 28.2.33 [pilot CW Bebb]; sold OR Guard 11.34

DH.82	G-ABUL	Regd 4.3.32; [pilot Turner Hughes] sold DH Acft Co Ltd 1/33 *[see below]*
HP Clive	G-ABYX	Regd 13.8.32; *"Youth of Britain"*; later *"Astra"*; regd Alan Cobham Avn Ltd 10.4.33; regd National Aviation Displays Ltd 27.11.33; regn cld 11.35

1932 Tour Aircraft – Other Participants

Avro 504K	G-ABHI	[Cornwall Avn Co] pilots Martin Hearn/Phillips
Avro 504K	G-AAUJ	[Cornwall Avn Co]; pilot Crundall; Fatal crash Harrogate 5.10.32
Avro 504K		[Cornwall Avn Co] [from G-EBIZ/EBYW/AAAF] pilot H Lawson
Avro 504K		[Cornwall Avn Co] added 6.32; pilot JD Parkinson
Avro 504K		[Surrey Flying Services] added 6.32
BAC VII Glider		[CH Lowe-Wylde]
Blackburn Bluebird		[unidentified and not operated 9.32]
Comper Swift	G-ABPY	[Brian Lewis & Co]; pilots Flt Lt Turner Hughes & Ogden [not operated 9.32]
Desoutter	G-AANE	[Rollason Muir & Rickard] pilot T Nash
DH.60G	G-ABJC	[DH Acft Co Ltd]
DH.80A	G-ABLB	[National Benzole Ltd] pilot JJ Scholes
DH.83		[Air Tours Ltd]
HP W.10	G-EBMR	[hired from Imperial A/w *"City of Pretoria"*] [pilot EB Fielden] [Not operated 9.32]
Southern Martlet	G-ABBN	[National Flying Services Ltd][pilot AC Rawson]; [Not operated 9.32]
Spartan 3-str	G-ABWV	[Rollason Muir & Rickard] pilot F Kent joined tour 17.6.32

Winter 1932/33 South African Tour – Owned Aircraft

Cierva C.19/IVP	G-ABGB	see above; Cr Cape Town 13.2.33
DH.66	G-ABMT	Regd Sir Alan Cobham 10.32 ex Imperial Airways *"City of Cape Town"*; sold back to Imperial Airways Ltd 3.33

Winter 1932/33 South African Tour – Other Participants

AW.16	G-ABKF	[Armstrong-Whitworth Acft *"Red Devil"*] pilot Turner-Hughes; Engine seized & returned to UK early
Avro Tutor	G-AARZ	[AV Roe & Co Ltd]
Avro Tutor	G-ABZP	[AV Roe & Co Ltd]; sold [National Avn Displays Ltd 3.35]
Avro Tutor	G-ABZR	[AV Roe & Co Ltd]; Fatal crash Cape Town 17.2.33 [Harold Lawson killed]

1933 Tour – Owned Aircraft

Airspeed Ferry	G-ABSI	see above; [pilot Joe King]
Airspeed Ferry	G-ABSJ	see above; [pilot CK Turner-Hughes/JD Parkinson]
Cierva C.19/IVP	G-ABUC	Regd Sir Alan Cobham 22.3.33; [pilot F Jaques]; sold FG London 8.36
Cierva C.19/IV	G-ABFZ	Regd Sir Alan Cobham 28.8.33 [pilot H Rawson]; sold HR Starkey-Howe 12.35
DH.60G	G-ABBX	Regd Sir Alan Cobham 7.4.33; [pilot B Wilson]; sold GE Archdale 9.34
DH.60G	G-ABUB	see above; [pilot R Warner]
DH.82	G-ABUL	Regd 4.33; regd Sir Alan Cobham 7.4.33; [pilot Geoffrey Tyson]; sold 11.33
DH.82	G-ACEZ	Regd 20.3.33; regd Sir Alan Cobham 7.4.33; [pilot CWH Bebb]; sold Brooklands Aviation Ltd 10.35
DH.82	G-ACFA	Regd 20.3.33; Not used and sold DH Acft Co Ltd 4.33
DH.83	G-ACEX	Regd 3.33 *"Youth of Ireland"*; regd Sir Alan Cobham 7.4.33; [pilot R Ogden]; sold Provincial Airways Ltd 5.34
DH.83	G-ACEY	Regd 3.33 *"Youth of Newfoundland"*; regd Sir Alan Cobham 7.4.33; [pilot Easdown]; sold Provincial Airways Ltd 5.34
HP W.10	G-EBMM	Regd Sir Alan Cobham 20.4.33 *"Youth of New Zealand"*; [pilot CH Bremridge]; regd National Aviation Displays Ltd 27.11.33; Fatal crash Aston Clinton 22.9.34
HP W.10	G-EBMR	Regd Sir Alan Cobham 20.4.33; [reserve acft]; regd National Aviation

		Displays Ltd 27.11.33; CofA lapsed 23.9.34 & scrapped Malta
HP.35 Clive	G-ABYX	see above; [pilot Flt Lt HC Johnson]
Lincock II	G-AALH	Regd Sir Alan Cobham 9.5.33; [pilot G Tyson]; regd National Aviation Displays td 27.11.33; CofA lapsed 27.3.35
Lincock III	G-ABFK	Regd Sir Alan Cobham 9.5.33; [pilot CWH Bebb]; regn cld 2.34 & donated to College of Aeronautical Engineering, Brooklands 6.34

1933 Tour Aircraft – Other Participants

Avro 504K	G-ABVH	[Air Travel Ltd] pilot George Kemp
Avro 504K	G-ABVY	[Air Travel Ltd] pilot W Miller
Avro 504K	G-ACCX	[Air Travel Ltd] pilot AN Kingwill
Avro 504K	G-EBIS	[L Rimmer owner/pilot]
Avro 504K	G-EBXA	[L Rimmer] pilot J Mackay
Avro 504K	G-EBGZ	[L Rimmer] pilot Jones
BAC Drone		
DH.83	G-ABWF	[Airwork Ltd] pilot G Tyson; Collided with DH.60G EI-AAI Limerick 7.7.33
Klemm 3-str		[Aerofilms Ltd] unconfirmed; replaced by Spartan 3-str
Klemm 3-str		[Aerofilms Ltd] unconfirmed; replaced by Spartan 3-str
Spartan 3-str	G-ABYH	[Aerofilms Ltd]
Spartan 3-str	G-ABZH	[Aerofilms Ltd]; Fatal cr Colpe Farm, Drogheda, IFS 1.10.33

NATIONAL AVIATION DISPLAYS LTD
Grand Buildings, Trafalgar Square, London WC2 [based Ford]
Formed 5.1.34 – for Third Annual National Aviation Day Display; commenced Dagenham 14.4.34 to 30.9.34 [163 towns]; directors – Sir Alan Cobham, Lady Gladys M Cobham, Edward M Rossiter [& publicity mgr]; Sidney T Morris;
Dallas L Eskell [genl mgr]; FTK Bulmore [announcer & pilot]
Pilots – Flt Lt Hugh C Johnson [chf pilot]; Charles WH Bebb; WV Creates; WH Easdown; JR King; P Phillips; Maj HG Travers; GV Tyson; Miss Joan Meakin; RJ Ashley; Flt Lt J Hyland; CH Bremridge; AN Woger-Slade; L Rowley; with Martin Hearn [wing-walker] and I Price & Naomi Heron-Maxwell [parachutists]
Avro Cadets G-ACLU/ACOZ/ACPB were painted red/white/blue & operated as *"Three Aces"* led by Geoffrey Tyson
Fourth [1935] Tour commenced Redhill 13.4.35 [additional pilots incl E Stewart, GE Collins, Leonard Carruthers]
Business and [some] assets sold to **CWA Scotts Flying Display Ltd** 12.35 [incl Ferry & 2 Cadets]

1934 Tour – Owned Aircraft

Airspeed Ferry	G-ABSI	see above [pilot Joe King]
Airspeed Ferry	G-ABSJ	see above
Avro 504N	G-ACOK	Regd 3.34; [pilot P Phillips]; sold LJ Rimmer 3.36
Avro 504N	G-ACRS	Regd 2.5.34; [pilot Mackay]; Fatal crash Cove, nr Farnborough 30.6.34
Avro 640 Cadet	G-ACLU	Regd Sir Alan Cobham 11.33; [pilot Easdown]; regd National Aviation Displays Ltd 2.4.35; sold CWA Scotts Flying Display Ltd 3.36
Avro 640 Cadet	G-ACOZ	Regd Sir Alan Cobham 3.34; [pilot Bebb]; regd National Aviation Displays Ltd 2.4.35; sold CWA Scotts Flying Display Ltd 3.36
Avro 640 Cadet	G-ACPB	Regd Sir Alan Cobham 3.34 [pilot HG Travers]; regd National Aviation Displays Ltd 2.4.35; sold CWA Scotts Flying Display Ltd 3.36
Cierva C.19	G-ABUC	see above [pilot RJ Ashley] [possibly also G-ABFZ]
DH.82	G-ACEZ	see above [pilot Tyson]
HP W.10	G-EBMM	see above [pilot HC Bremridge]
HP W.10	G-EBMR	see above [reserve acft]
HP.35 Clive	G-ABYX	see above [pilot HC Johnson]
Lincock II	G-AALH	see above [pilot GV Tyson]

1934 Tour Acft – Other Participants

Rhonbussard		[pilot Miss Joan Meakin]
Spartan 3-str	G-ACAF	[pilot Capt EW Jordan]
Avro 504N	G-ACPV	[LG Anderson] [pilot J Hyland]

1935 Tour – All Owned Aircraft

Airspeed Ferry	G-ABSI	see above [pilot JR King]
Avro 504N	G-ACLV	Regd 1.3.35; sold Aerial Sites Ltd 10.35
Avro 504N	G-ACOD	Regd 1.3.35; [pilot E Stewart]; fatal collision Wessex G-ADFZ Blackpool 7.9.35
Avro 504N	G-ACPV	Regd 2.4.35; sold Aerial Sites Ltd 10.35
Avro 504N	G-ADBD	Regd 1.6.35; sold Air Travel Ltd 11.35
Avro 621 Tutor	G-ABZP	Regd 28.3.35; [pilot Tyson]; sold North Eastern Airways Ltd 6.37
Avro 640 Cadet	G-ACLU	see above [pilot Easdown]
Avro 640 Cadet	G-ACOZ	see above [pilot Bebb]
Avro 640 Cadet	G-ACPB	see above [pilot Travers]
Cierva C.30A	G-ACYH	Regd 4.5.35 [pilot RJ Ashley]; sold Aerial Sites Ltd 2.36
DH.82	G-ACEZ	see above [pilot G Tyson]
Grunau Baby 2		Owned Sir Alan Cobham; [1st prodn acft bought from Slingsby]; [pilot GE Collins]; Fatal crash Upwood, Ramsey, Huntingdon 30.7.35
HP Clive	G-ABYX	see above [pilot HC Johnson]
Pou du Ciel	G-ADSC	Regd 9.35 [built Felix Louis]; sold Scott Motors [Saltaire] Ltd 11.35
Pou du Ciel	G-ADSD	Regd 9.35 [built Felix Louis; HM14 No.4]; sold WG Barnett 11.35
Pou du Ciel	-	#3 not completed; sold to Gilberts
Wessex	G-EBXK	Regd 29.7.35; CofA lapsed 3.5.36
Wessex	G-AAGW	Leased from Imperial Airways 1.7.35; Damaged Wick 7.35
Wessex	G-ADFZ	Regd Sir Alan Cobham 4.35; [pilot Carruthers]

NAVARRO AVIATION CO LTD
Burton-on-Trent [2/19]; Southsea Road, Kingston-on-Thames [12.20; 7.21]
Formed by Joseph G Navarro – Joyriding Whitstable/Southend 1919

Avro 504K	G-EADY	Regd 6.6.19 [E3408]; sold Astra Eng'g Works & Garage 5.20
Avro 504K	G-EAEA	Regd 6.6.19 [D9304]; sold Astra Eng'g Works & Garage 5.20
Avro 504K	G-EAEB	Regd 6.6.19 [B8774]; sold Astra Eng'g Works & Garage 5.20
Avro 504K	G-EAJP	Regd 8.19; sold Astra Eng'g Works & Garage 5.20

NORTH BRITISH AVIATION CO LTD
Hooton Park Aerodrome, Liverpool, Cheshire
Formed 2.3.29 by Lance J Rimmer [pilot, ex Berkshire Avn], EE Fresson [sold out late 1932] and W Mackay

Avro 504K	G-EBIS	Regd 27.2.29; regn cld 12.34 [but reported as Crashed Rainhill 22.4.35]
Avro 504K	G-EBXA	Regd 27.2.29; CofA lapsed 18.12.32
Avro 504K	G-EBGZ	bt 26.7.29; regd 7.10.29; CofA lapsed 17.2.34
Avro 504K	G-EBSJ	Regd 11.31; sold GH Lloyd 5.34

THE NORTHERN AVIATION CO
Royal London Buildings, 196 Deansgate, Manchester; to 186 Oxford Rd, Manchester [3.24][based Alexandra Park, Manchester – closed 10.24]

Avro 504K	G-EAZW	Regd 17.3.24; Crashed Cleckheaton [or Brighouse], nr Bradford 3.5.24
Avro 504K	G-EAZX	Regd 17.3.24; Crashed Manchester 2.8.24
Avro 504K	G-EADP	Regd 19.3.24; Crashed Morecambe Foreshore 27.7.26
DH.9	G-EBDG	Regd 3.24; sold Wm Beardmore & Co Ltd 7.24

NORTH SEA AERIAL & GENERAL TRANSPORT LTD
Olympia, Roundhay Rd, Leeds; later Brough, East Yorkshire [based Brough]
Formed 23.4.19 as **THE NORTH SEA AERIAL NAVIGATION CO LTD** as subsidiary of **Blackburn Aeroplane & Motor Co Ltd**; renamed 18.10.19
Joy-riding activities involved:

Blackburn Kangaroo	G-EAIT	Regd 1.8.19 [B9978]; CofA lapsed 10.8.21
Blackburn Kangaroo	G-EAIU	Regd 1.8.19 [B9973]; CofA lapsed 19.4.29

Blackburn Kangaroo	G-EAKQ	Regd 18.8.19 [B9972]; sold Peruvian Centro Militar, Las Palmas 20.21
Avro 504K	G-EAGV	Regd 14.7.19 [H6598]; sold Olivers Mount, Scarborough 3.8.20
Avro 504K	G-EAGW	Regd 14.7.19 [H6599]; Crashed Scarborough [7.20]

OXFORDSHIRE AVIATION CO
12 Woodcote Rd, Caversham [based Caversham, Reading]
Founded 1920 by Capt JM Drysdale & Mr Gomez-Cornejo and undertook joy-riding at Newport, Mon [G-EADU] 6.20-11.20; re-commenced operations 18.2.21

Avro 504K	G-EAGT	Regd 9.8.20; cld 3.21
Avro 504K	G-EADU	Regd JM Drysdale 11.9.20; sold DM Mathews 8.23

PLEASURE FLYING SERVICES
Cramlington Aerodrome
Owned by Constance Leathart & WL Runciman and taken over by **Cramlington Aircraft Ltd** 9.29

Avro 548	G-EBPO	Op 12.28; Regd 2.29; sold Cramlington Acft Ltd 10.29
Spartan 3-str	G-AAGV	Regd 4.29; sold Cramlington Acft Ltd 10.29
Spartan 3-str	G-AAHV	Regd 8.29; sold Cramlington Acft Ltd 10.29

POPULAR FLIGHTS LTD
1 Albemarle St, London W1 [based Somerton, Cowes, IoW]
Formed 18.7.29 by GS Hewitt [MD], PH Davies, Miss VET Latham

Avro 504K	G-AAEM	Regd 24.7.29; Crashed Somerton 16.8.29

PSYCHE'S FLYING CIRCUS
Trading name for operations by **Herbert Sykes**, 23 Christchurch Ave, Brondesbury, London NW6 [based Kingsbury, Middlesex]

Avro 504K	G-EANN	Regd 3.20; sold Adastral Air Lines Ltd 9.20
Avro 504K	G-EACA	Regd 8.2.21; Crashed Earlham Road, Norwich 28.9.21
Avro 504K	G-EBAY	Regd 1.22; No CofA; sold A Fraser 7.22
Sopwith Pup	G-EBAZ	Regd 1.22; No CofA; sold PF Capon 6.23
Avro 504K	G-EBBF	Regd 1.22; No CofA; sold A Fraser 7.22

THE SCOTTISH MOTOR TRACTION CO LTD
29 East Fountainbridge, Edinburgh 3 [based Renfrew; Turnhouse 5.33; Macmerry wef 6.34]
Scottish coach business which formed airline 7.32 to operate private charter services and joyriding
Chairman/MD - Bailie [later Sir] William J Thomson [also Lord Provost/Lord Lieutenant of City of Edinburgh]
Managed by John C Sword [Dir & Manager of Western SMT following merger early 1930s] - see Midland & Scottish Air Ferries Ltd
Pilots: Sqn Ldr Henry GR Malet [chief pilot; appointed Aviation Manager 3.33]; Flt Lt NMS Russell [appt Air Superintendent 4.33]; WB [Bill] Caldwell [chief pilot 4.33]; Capt GB Holmes

DH.83	G-ABWB	Regd 7.32; sold North Sea Aerial & General Transport Ltd 7.33
DH.83	G-ABWF	Regd 7.32; Crashed Helmshore, Lancs 31.1.33 [to Airwork Ltd 4.33 for rebuild]
DH.60G	G-AADA	Regd 6.3.33; sold L Russell 5.33
DH.83	G-ACDZ	Regd 15.3.33; sold West of Scotland Air Services 11.34
DH.83	G-ACEA	Regd 15.3.33; sold Sandown & Shanklin Flying Services Ltd 7.36
DH.83	G-ACEB	Regd 15.3.33; sold Southend Flying Services Ltd 5.34
DH.83	G-ACEC	Regd 15.3.33; sold West of Scotland Air Services 4.35
Avro 640 Cadet	G-ACFH	Regd 18.3.33; CofA lapsed 27.4.34
DH.82	G-ACDY	Regd 20.3.33; Crashed into River Den, Aberdeen 26.8.33
DH.83	G-ACED	Regd 12.4.33; sold Northern & Scottish Airways Ltd 6.35
DH.83	G-ACEE	Regd 12.4.33; Fatal crash Riverside Park, Dundee 31.7.34
Avro 640 Cadet	G-ACFS	Regd 12.4.33; CofA lapsed 17.5.34
Avro 640 Cadet	G-ACFT	Regd 12.4.33; sold Aircraft Facilities Ltd 6.35
Avro 640 Cadet	G-ACFU	Regd 12.4.33; sold T Campbell Black 3.35
DH.84	G-ACDM	Regd 21.4.33; sold to ZS-AEI 3.34
DH.84	G-ACDN	Regd 21.4.33; to MSAF 22.9.33
DH.83	G-ACEI	Regd 21.4.33; Crashed/dbf Alva, Clackmannanshire 1.7.33
DH.83	G-ACEJ	Regd 21.4.33; sold Norfolk & Norwich Aero Club Ltd 4.35
DH.84	G-ACET	Regd 21.4.33; sold MSAF 2.34
DH.84	G-ACRI	Regd 5.34; sold Lt Col NA Blandford-Newson 9.34

SKY DEVILS AIR CIRCUS
Trading name for **Air Pageants Ltd** [4.34]

SKYWORK LTD
Ashley House, 185 Fleet St, London EC [based South Africa]
Formed 23.9.31 by John Tranum [parachutist] and Oscar Garden [of Vacuum Oil]
Pilots [mainly ex CD Barnard Air Tours Ltd]: ED Ayre, Oscar Garden; CEF Reilly, ED Cummings, JR [Joe] King
Op South African flying circus [sponsored by Vacuum Oil Co of S Africa] *"The Spartan Air Circus"*, aircraft dep UK 9.10.31; tour commenced 10.31, completed 2.32

Spartan 3-str	G-ABPZ	Regd 30.9.31; regn cld 11.32 – to ZS-ADP
Spartan 3-str	G-ABRA	Regd 30.9.31; sold J Stark 10.32
Spartan 3-str	G-ABRB	Regd 30.9.31 *"Miss Mobiloil"*; regn cld 11.32 – to VR-TAJ 10.32
Desoutter	G-AAPP	Regd JR King 15.10.31; Crashed Cape Town 21.11.31

SOUTHERN COUNTIES AVIATION CO [LTD]
Port Meadow, Oxford; to Brooklands Aerodrome, Byfleet [10.24] – joyriding in IoW 1925
Formed 1924 by Geoffrey V Peck, JPC Phillips and Christopher PB Ogilvie [Coy formed 1925]
Pilots: Thomas K Breakell [8.24; 10.24]; CWH Bebb; Ronald H Leavey [5.26 – killed in G-EASG 2.9.26]; Leslie A Lewis [5.26; 4.27]; Mrs SC [Mary] Elliott-Lynn [Summer 1926]
One unidentified aircraft crashed in garden Addlestone 24.12.25; 5 chickens killed

Avro 504K	G-EBJE	Regd GV Peck 7.24; sold JR Cobb 1.26
Avro 504K	G-EASG	Regd 6.5.25; Fatal crash nr Great Bookham 2.9.26
Avro 548	G-EBAG	Regd 5.25; sold British Avn Insurance Group 8.26
Avro 504K	G-EBKS	Regd 8.6.25 *"Attaboy"*; CofA lapsed 9.6.27
Sopwith Gnu	G-EAGP	Regd 11.7.25; sold Lloyds Commercial Acft Co 1926
Sopwith Gnu	G-EADB	Regd 8.25; sold Lloyds Commercial Acft Co 2.26
Avro 504K	G-EAAY	Regd 23.4.26 *"Jake"*; sold FG Miles 10.26

SOUTH WALES AIRWAYS [LTD]
The Bridge End, Glamorgan [based strip at Coity, 1.5 ml NE of Bridgend]; 9 Dunraven Place, Bridgend, Glamorgan [1.33]; Wenvoe Aerodrome, Cardiff [7.34]
Formed 5.27 by Robert H Thomas [MD], DW Griffiths and LJ Rimmer [Company formed 4.3.31]

Avro 504K	G-EBNH	Regd 3.5.27; Crashed Bridgend 7.5.28
Avro 504K	G-EBSG	Regd 11.7.27; sold Western Airways Ltd 5.28
Avro 504K	G-ABLV	Regd RH Thomas 5.31; CofA lapsed 15.5.32
Avro 504K	G-ABLW	Regd RH Thomas 5.31; No CofA issued; sold JHA Wells 4.33

S SUMMERFIELD & CO
17 Craven St, Melton Mowbray; later The Park, Melton Mowbray
Prop: Samuel Summerfield – joyriding on East Coast/Yorkshire

Avro 504K	G-EADR	Regd 9.3.20; Crashed Werrington, Northants 24.6.20
Avro 504K	G-EAEB	Regd 26.7.20; Crashed Norwich 25.9.21
Avro 504K	G-EASA	Regd 14.11.21; Crashed Doncaster 28.4.22

STALLARD AIRWAYS
Parkfield, Somerfield Rd, Sutton Coldfield [based Herne Bay/Penshurst]
Formed 1920 by Mrs JS Stallard

Pilots: Ronald H Leavey [21/22 – co-owner of business?]; Tommy Baden-Powell [joined 3.22]; Arthur Boorman

DH.6	G-EANJ	Regd JS Stallard 19.7.20; CofA lapsed 18.4.22
Avro 504K	G-EAFQ	Regd 2.9.21; Crashed Hildenborough/Penshurst, Kent 26.11.21
Avro 504K	G-EAJG	Owned T Baden-Powell 21.22; Crashed Penshurst 20.8.22
Avro 504K	G-EASK	Owned A Boorman 21.22

SURREY FLYING SERVICES [LTD]
Croydon Aerodrome [also joy-riding initially at Castle Bromwich; pilot Haydon]
Formed 2.21 by FWJ Grant and Capt AF Muir for joy-riding/air taxi & flying training [Company formed 5.9.31]
Robert & Graham Little [joined 1923 ex Border Aviation]
Pilots: Capt AF Muir [2.21; 8.21]; Jimmy Youell [12.22]; Haydon [appt 2.21; rejoined RAF 4.21; returned 4.23]; JC Chamberlain; GLP Henderson [5.26]; EF Smith [5.26; killed in cr G-EBZD Wallington 2.4.31]; JJ Flynn [5.26, 6.29]; Sydney F 'Timber' Woods [chf inst; to Rollasons 3.32; ret; dep 6.36]; McConnel [9.27]; Malcolm A Cowan [appt chf pilot 1931]; Joe C Chamberlain [pilot – joyflights]

Avro 504K	G-EAWI	Regd 25.2.21; Crashed nr Croydon 9.21 [probably the 504K dgd Swanley late 8.21?]
Avro 504K	G-EAWJ	Regd 25.2.21; regn cld 28.2.22
Avro 504K	G-EAIR	bt 8.21; Regd 2.9.21; Crashed Hayling Island 11.8.23
Avro 504K	G-EBDP	Regd 22.6.22; CofA lapsed 24.7.25
Avro 548	G-EBBC	Regd 2.8.22; CofA lapsed 26.8.24
Avro 504K	G-EBFW	Regd 1.5.23; Crashed Mudford, Somerset 17.9.26
Avro 504K	G-EBHM	Regd 30.7.23; Crashed/dbf Port Talbot 31.5.27
Avro 504K	G-EBII	Regd 15.10.23; CofA lapsed 5.11.24
Avro 536	G-EAKJ	Regd 22.4.24; CofA lapsed 16.5.26
Avro 536	G-EAKP	Regd 30.6.25; CofA lapsed 14.7.26
Avro 536	G-EAKM	Regd 30.7.25; Crashed Taplow, Bucks 4.7.28
Avro 536	G-EBOF	Regd 31.5.26; CofA lapsed 5.3.29
Avro 536	G-EBOY	Regd 23.7.26; CofA lapsed 11.4.29
Avro 536	G-EBRB	Regd 27.4.27; Crashed Barry 28.5.28
Avro 536	G-EBTF	Regd 8.27; CofA lapsed 31.8.28
Avro 548A	G-EBIV	Regd 5.28; sold EG Clerk 3.30
Avro 504K	G-EBYW	Regd 7.28; sold Aviation Tours Ltd 5.30
Avro 504K	G-EBZB	Regd 7.28; sold MW Allenby 11.32
Avro 548	G-AABW	Regd 17.9.28; Fatal crash Sidcup Bypass, nr Welling 19.9.29
Avro 504K	G-AAGB	Regd 4.29; sold Universal Airways 7.34
Avro 504K	G-AAYM	Regd 5.30; sold Devonshire Air Tours 4.33
Avro 504K	G-ABAY	Regd 3.6.30; regn cld as sold 1.36
Avro 504K	G-ABZC	Regd 9.32; sold Devonshire Air Tours 4.33
Avro 504K	G-AAAF	Regd 16.8.33; sold HV Kimberley Atkinson 6.35

THANET AVIATION LTD
17 Chapel Place, Ramsgate, Kent [based/op Nethercourt Aerodrome [Cheesman's Farm], Ramsgate]
Formed 25.2.31 by Percy Turner, Ernie Bicknell [pilot], JT Huddlestone, JL Barnes
Op joyrides 7.31

| Avro 504K | G-ABJF | Regd 11.3.32; CofA lapsed 16.5.33 |

UNIVERSAL AIRCRAFT SERVICES LTD
6 Broad St Place, London EC2; to Witney Airfield, Minster Lovell, Oxon [based Witney; also at The Rock Hotel, Llandrindod Wells wef 8.33]
Formed 21.12.32 to incorporate business of Donald L Townsend [MD] and HBG Micklemore
Pilots – F/O Douglas FC Brecknell; EMH Slade

| Bristol Fighter | G-ACCG | Regd 2.33; sold MN Mavrogordato 10.33 |
| Avro 504K | G-ABWK | Regd 29.4.33; sold EG Clerk 8.34 |

UNIVERSAL AIRWAYS
20 Bride Lane, Fleet St, London EC4 [based Telscombe Cliffs, Newhaven, Sussex]

| Avro 504K | G-AAGB | Regd 7.7.34; Crashed Telscombe 11.7.34 |

VULCAN AVIATION CO LTD
Hull [operated Sands Lane Estate, Bridlington]
Formed 1921 by Charles Smith

Pilots: Capt AE Freitag; Capt EH Smith
Ceased ops by 8.21

| Avro 504K | G-EAJK | Regd 24.5.21; CofA lapsed 21.8.21 [possibly after forced landing nr Darlington 6.21?] |

WARWICK AVIATION CO

| DH.6 | G-EAHH | Regd 18.7.19 [F3435]; CofA lapsed 10.8.20 |
| DH.6 | G-EAHI | Regd 18.7.19 [C6889]; sold Manchester Avn Co 10.20 |

THE WELSH AVIATION CO LTD
31 Fisher St, Swansea [based West Cross, Swansea; also Blackpill, Swansea Sands]
Formed 11.20 by FGM Sparks [MD] & Capt EA Sullock
HS Broad [appt pilot 4.21]
Ceased ops 1922 & sold to turf accountant Evan Williams, Port Talbot

Avro 504K	G-EAWK	Regd 3.21; sold E Williams 4.22
Avro 504K	G-EAWL	Regd 3.21; sold E Williams 4.22
Avro 504K	G-EAWM	Regd 3.21; sold E Williams 4.22
Avro 548	G-EAFH	Regd 5.21; sold ED Whitehead Reid 3.22

WESTERN AVIATION LTD
1 Leamington Place, Cheltenham [based Cheltenham]
Formed 26.3.27 by Capt Edward W Jordan [ex Cornwall Avn Co], J Sheils
Joy-riding coy; ops commenced in West Midlands 12.4.27
Taken over by **Aviation Tours Ltd** 1933

Avro 504K	G-EBQR	Regd 4.27; sold Aviation Tours Ltd 4.33
Avro 504K	G-EBXV	Regd 4.28; sold Aviation Tours Ltd 4.33
Avro 504K	G-EBSG	bt 5.28 [not regd]; CofA lapsed 28.7.28

WEST OF SCOTLAND AVIATION CO LTD
Crawford St, Glasgow [based Renfrew]
Formed 5.8.19 by TH French, FW French, RW Reeve, J Fryer

Avro 504K	G-EAHY	Regd 8.19; CofA lapsed 15.8.20; sold Wm Beardmore & Co Ltd 9.23
Avro 504K	G-EAHZ	Regd 8.19; CofA lapsed 15.8.20; sold FJV Holmes 8.21
Avro 504K	G-EAIA	Regd 8.19; sold Border Avn Co Ltd 8.20
Avro 504K	G-EADH	Regd 10.19; No CofA issued; sold Kingwill & Jones Flying Co 5.22

WILLIAMS & CO
5 Sandgate, South Shore, Blackpool, Lancs [4.34]; 117 Squires Gate Lane, Blackpool 7.29 [based Squires Gate]
Formed 1934 by Leslie CL Williams
LA Lewis [pilot 4.38 – fired]
T/a **Monarch Airways** 1939 [based Stanley Park?]

Avro 548	G-EBOK	Regd 14.4.34; CofA lapsed 20.3.35
Avro 548A	G-EBIU	Regd 3.10.34; CofA lapsed 16.5.37 [& dbr Rhyl 17.5.37?]
Avro 504K	G-ABAA	Regd 12.5.36; sold Flt Lt Birch 9.38
GAL ST.25	G-ADYN	Regd 17.8.38; Impressed as X9373 31.3.40
Short Scion	G-ADDP	Regd 4.7.39; Impressed as X9374 31.3.40

WOLVERHAMPTON AIRCRAFT CO LTD
Waterloo House, 20 Waterloo St, Birmingham [based Wolverhampton]
Formed 6.1.30 by JR Whitehill, C Wrighton

| Avro 504K | G-AATY | Regd 22.1.30; CofA lapsed 8.5.34 |
| Avro 504K | G-EBKR | Regd 29.5.30; Crashed Church Lawton, Cheshire 22.7.32 |

ZENITH AIRWAYS LTD
Camber Aerodrome, Camber, Sussex [based Camber]
Formed 3.5.35 by Herbert D Ward, George T Butler
Possibly connected with **Aviation Commerce Ltd**

| Avro 504N | G-ADGB | Regd 6.6.35; regn cld 6.36 |
| Avro 504N | G-ADGC | Regd 6.35; No CofA issued; sold JE Coxon 11.37 |

APPENDIX III
Dramatis Personae

Many individuals were associated with the joy-riding, pleasure flight and touring air displays during the inter-war period. As with the companies, little information is now available on some, yet a great deal on others. Presented here as 'Personality Profiles' are those individuals generally considered to have played pre-eminent roles during this entertaining period in British aviation.

Little is known of the history of Bernard Martin, whose Martin Aviation Company operated DH.6s as seen here.

Allison, Percy

Allison was first engaged by Sir Alan Cobham as a ground engineer for his 1932 Air Display tour. Naturally outspoken and not prepared to put up with a boss who frequently required to 'let off steam', he found himself sacked on more than one occasion. However, by keeping a low profile he managed to survive and gained Cobham's confidence such that he was sent to India as part of a small team to evaluate the possibility of a touring display. He also learned to fly the Cierva Autogiro, then a complete rotary-winged novelty. Allison stayed in Cobham's employ until retirement in 1969, becoming a highly valued practical engineer at the forefront of air-to-air refuelling development. He died on 19th October 1996.

Anderson, Leslie George

George Anderson flew with the British Hospitals Air Pageant in 1933 but was also a spares/second-hand aircraft dealer at Hounslow with his brother Lionel John. Founder of Jubilee Air Displays Limited in February 1935, he was killed in the crash of the company's Avro 504N G-ADBS at Bodmin on 15th August 1935 during a joy-ride flight.

Ayre, Edward Dundonald

Born 2nd March 1900, 'Don' Ayre flew with a succession of companies including Imperial Airways, Henderson School of Flying, Lancashire School of Aviation and National Flying Services, where he was the manager of Hull Aero Club. His joy-riding activities were undertaken with CD Barnard Air Tours Ltd in 1931, and Skywork Ltd (Spartan Air Circus) in South Africa 1931-32, flying Desoutter G-AAPP. He was later a pilot for BAC Ltd, testing their Drones in 1932 and became workshop manager for Airwork Ltd in 1933 He was appointed manager of Dyce Aerodrome in 1935.

Barker, Henry ACG

Tantalisingly few personal details are available of Harry Barker, a business entrepreneur who, with partner James McEwan King, promoted several touring air displays in the early 1930s. Originally engaged as secretary for Sir Alan Cobham's National Aviation Day display in 1932, he 'defected' to form the rival Air Pageants Ltd which traded as the British Hospitals Air Pageant the following year. Despite severe public criticism regarding the amount of monies actually donated to the hospitals, Barker (and McEwan King) formed a succession of touring organisations that featured the top performers on the exhibition circuit. These included the Sky Devils Air Circus (1934), Jubilee Air Displays (1935), British Empire Air Displays (1936) and Aircraft Demonstrations Ltd which traded as Coronation Air Displays (1937). With the decline of public interest in air shows, the latter company was forced into liquidation late in 1937. At the formal hearing, the reasons given were omnibus strikes, bad weather and a boycott of the displays following border outrages at the time of the Royal visit to Ulster.

Barnard, Charles Douglas

After wartime service Barnard became a test pilot for the Sopwith Aviation Company and, in 1920, the de Havilland Aircraft Company Ltd, where he also later flew as an instructor for the School of Flying. He became the Duchess of Bedford's private pilot in 1927 and flying Fokker F.VII G-EBTS *"Princess Xenia"*, he carried out record flights with her to India and South Africa. In March 1930, Barnard formed aircraft dealers Brian Lewis & CD Barnard Ltd at Heston, departing when it was sold to Selfridges in August 1931. He formed CD Barnard Air Tours Ltd in March 1931, which led to the promotion of 'The Barnardstormers' tour of the UK between April-October 1931. This was succeeded by India Air Pageants Ltd which he operated on the sub-continent between November 1933 and April 1934. Barnard entered the RAF in 1940, becoming personal pilot to a high-ranking officer, and joined Flight Refuelling Ltd at Staverton in 1942 as an experimental pilot before retiring in 1944.

Bebb, Charles William Henry

'Cecil' Bebb, as he was universally known, was born on 27th September 1905. After wartime service he became a pilot for Southern Aircraft Ltd at Shoreham in 1931. He then became a star attraction with Sir Alan Cobham's National Aviation Day tours, including the visit to South Africa in 1932-33. Bebb continued to fly with Cobham Air Routes Ltd, a short-lived airline formed to operate between Britain's south coast and the Channel Islands, before joining Olley Air Service Ltd when Gordon Olley took the operation over in 1935. Bebb was the pilot who flew General Franco in Dragon Rapide G-ACYR to Spanish Morocco in July 1936 in the events leading up to the Spanish Civil War. He became chief test pilot at Cunliffe-Owen Aircraft Ltd during WWII but returned to Olley Air Service Ltd in 1946. After an adventurous flying career, Bebb, the last of the great flying entertainers engaged by Sir Alan Cobham died on 29th March 2002.

Bonar, Eric Watt

'Jock' Bonar, born in 1900, experienced his first taste of joy-riding as a pilot with the Lancashire School of Aviation at Blackpool in 1929. He subsequently became chief instructor at Northern Air Transport in 1932 and flew with the National Aviation Day display (1932), British Hospitals Air Pageant (1933) and the British Empire Air Display (1936). He acted as co-pilot to Colonel James Fitzmaurice in the Irish entry for the 1934 MacRobertson Race to Australia, the Bellanca EI-AAZ, *"Irish Swoop"*. In July 1937, Bonar was the test pilot for Deekay Aircraft Corporation Ltd at Broxbourne, Herts. After later serving as manager/instructor at Luton Flying Club he joined Rolls Royce in 1938.

Bruce (neé Petre), The Hon Mrs (Mildred Mary) Victor

Born on 10th November 1895, Mrs Bruce achieved celebrity following her diverse land, sea and air sporting activities during the inter-war period. Her achievements in the air embraced a round-the-world solo flight in a Blackburn Bluebird G-ABDS that included the first solo flight from

England to Japan in 1930. In 1932, she conducted air-to-air refuelling experiments with the Saro Windhover flying boat G-ABJP and specially converted Bristol Fighters for an attempt on the woman's world endurance record. Though unsuccessful in achieving any records, her in-flight refuelling initiative pre-empted that of Sir Alan Cobham whose commercial endeavours later proved successful. Mrs Bruce formed Luxury Air Tours Ltd in 1932 and joined the British Hospitals Air Pageant tour of the United Kingdom the following year. Amongst her other ventures were the operation of an aerodrome at Hook under the auspices of the Thames Valley Aero Club (1933) and several charter and freight airlines including, in 1934, Air Dispatch Ltd and Commercial Air Hire Ltd. Her life in the 'fast lane' extended well into old age and she died after having well exceeded her *"Nine Lives Plus"* (autobiography title) on 21st May 1990.

Campbell Black, Thomas

Tom Campbell Black, born in 1899, belonged to that elite band of inter-war fliers whose names attracted instant international recognition. Having learned to fly in the Royal Naval Air Service in 1917, 'TCB' became well-known in the 1920s as an East African bush pilot and outstanding flying instructor. Beryl Markham, the first woman to fly single-handed across the Atlantic from east to west on 4-5th September 1936, became his most famous pupil. With two partners, in 1928 he purchased the DH.51 G-EBIR, *"Miss Kenya"* (now preserved at Old Warden), and the following year helped set up Wilson Airways Ltd which operated over largely uncharted East African territory where forced landings frequently spelled disaster. Returning to England in 1932, he later became the private pilot for Lord Furness before becoming a headline celebrity when, along with CWA Scott, he was first across the finishing line at Melbourne in the 1934 MacRobertson air race, flying one of three specially built DH.88 Comets G-ACSS, *"Grosvenor House"*. He married the actress Florence Desmond in March 1935 and set up Campbell Black (Aviation) Ltd the same month. In August 1935, he narrowly escaped with his life on an attempt to beat the England – Cape Town record when he and co-pilot J G MacArthur had to parachute from DH Comet, G-ADEF following the failure of both engines over the Sudan. In 1936, he was co-opted by Barker and McEwan King to head up their British Empire Air Display. As with a number of other pilots, he freelanced for the Spanish Nationalists in July and August 1936, prior to being sponsored by John Moores of Littlewoods Football Pools for the Schlesinger Air Race to Cape Town due to depart in late September. His life came to a tragic end however when, at a publicity event at Liverpool on 19th September 1936, he was fatally injured when his Percival Mew Gull, G-AEKL *"Miss Liverpool"* was struck by an RAF Hawker Hart, K3044.

Cathcart-Jones, Owen

Born on 5th June 1900, Cathcart-Jones served in the Royal Marines before flying for National Flying Services in 1930. He then became private pilot to Lt Cdr Glen Kidston and accompanied him in setting a new London – Cape Town record in 1931. In the 1934 MacRobertson air race, he flew as co-pilot with Ken Waller in Bernard Rubin's DH Comet G-ACSR, finishing fourth. Cathcart-Jones' entry into the world of air displays took place when he was appointed chief pilot for Barker and McEwan King's Jubilee Air Displays in 1935. In July 1936, he flew for the Spanish Nationalists, undertaking several clandestine missions between General Mola's headquarters in Spain and the exiled King Alfonso XIII in Czechoslovakia.

Cobham, Sir Alan John

Born on 6th May 1894, Alan Cobham learned to fly in the Royal Flying Corps. Along with the Holmes brothers he co-founded Berkshire Aviation Co in 1919, aka Cobham and Holmes Aviation Co, giving joy-riding flights throughout England and Scotland. After a short spell as an aerial photography pilot with the Aircraft Manufacturing Co Ltd (Airco), he spent six years 1921-27, as chief pilot with its successor, the de Havilland Aircraft Co Ltd. During this period he undertook long distance charter flights including, in 1923, a 12,000 mile tour of Europe, the Middle East and North Africa with a single passenger (Lucien Sharpe). A series of epic route-proving flights to India and Burma (1924), South Africa (1925) and, most notably, Australia (1926) brought Cobham international recognition and a knighthood.

By now three times winner of the Royal Aero Club's Britannia Trophy (1923, 1925 and 1926) and having flown DH.50 G-EBFN to victory in the 1924 King's Cup Air Race, Cobham left de Havillands to form Alan Cobham Aviation Co Ltd in 1927. A 20,000 mile exploratory flight around the African coast in Short Singapore G-EBUP led, in 1928, to the co-founding with Robert Blackburn, of Cobham-Blackburn Air Lines, but the proposed route structure was taken over by Imperial Airways and the company wound up in July 1930. In 1929, Cobham also became a founding director of National Flying Services Ltd and in that year carried out his Municipal Aerodrome Campaign in which, over a 21-week period, flying a ten-seat DH.61 Giant Moth G-AAEV *"Youth of Britain"*, he visited 110 towns in Great Britain and carried 50,000 passengers.

In March 1931, he bought a financial stake in and became a director of newly-formed aircraft manufacturer Airspeed Ltd and in July/August of that year, undertook another survey flight to Africa, this time in a Short Valetta floatplane, G-AAJY on behalf of Imperial Airways. Upon his return, Cobham started to plan his National Aviation Day tour for the following year, aimed at making the British public 'air-minded'. His determination resulted in successive tours of the United Kingdom over the years 1932-35, and a winter tour of South Africa 1932-33. Co-incident with his touring preparations, Cobham also in 1933 became involved with Gibraltar Airways Ltd, a small airline financed by the Bland Steamship Line to fly between Gibraltar and Tangier.

He eventually sold his touring and display organisation to CWA Scott in 1935, by which time he had founded, in October 1934, Flight Refuelling Ltd to investigate air-to-air refuelling techniques and equipment. Cobham, in addition to assuming a directorship of aircraft dealers RK Dundas Ltd, also attempted, albeit unsuccessfully, to operate a passenger service between Croydon, the South Coast and Guernsey under the name Cobham Air Routes Ltd which commenced operations in May 1935, but ceased trading two months later after the fatal crash of Wessex G-ADEW. Cobham had also became chairman of the Portsmouth, Southsea & Isle of Wight Aviation Ltd in March 1935, when the company was refinanced with the intention of co-ordinating its activities with those of his own airline, as well as supplying them with Airspeed Envoys and Couriers. His list of directorships was later extended when, in August 1937, he became a founding member of Crouch Bolas Ltd.

Meanwhile, in September 1934, he had financed an abortive long-distance non-stop flight to India with Airspeed Courier G-ABXN using air-to-air refuelling, and now finally abandoned his airline ambitions to concentrate upon developing this business. Despite success and disappointments in equal measure, Flight Refuelling Ltd went on to eventually become the world's leading supplier of airborne refuelling equipment and the cornerstone company of what is now Cobham plc. Sir Alan Cobham, KBE, AFC remained in charge of a rapidly growing organisation for thirty years before his death on 21st October 1973. In a smooth carry-over of family interests, Michael Cobham, (now Sir Michael), continued to exercise executive control for an identical period until his own retirement in 1995.

Crossley, Mrs Winifred Mary

Born at St Neots on 9th January 1906, Winifred Crossley 'gained her wings' with the Norfolk and Norwich Aero Club on 16th February 1934 and later the same year bought her own Gipsy Moth G-AAET. She subsequently made an impact on the air display circuit, being featured on CWA Scott's Flying Display advertising posters as 'Britain's only female aerobatic pilot' – and she had a medically approved certificate which supported her claim that a Guinness a day helped her to perform! Working as a contract pilot, she became a key member of Captain Phillips' Air Publicity banner-towing team in the mid-1930s. On one occasion following an engine failure over Wembley, she jettisoned the banner and glided down to land on a golf course. Apparently the golfers were suitably impressed and welcomed their unexpected guest with tea and brandy. Winifred Crossley went on to become one of eight pilots that formed the Air Transport Auxiliary (ATA) on 1st January 1940, re-marrying and becoming Mrs Fair during that period. She ferried hundreds of aircraft of every type throughout the United Kingdom until the disbandment of the ATA in 1945.

Davis, Henry Duncan

Duncan Davis was born on 3rd April 1896 and his colourful career in aviation began when he ran away from school to become Colonel Cody's first apprentice at Farnborough before joining Sopwith Aviation. Initially joining the Royal Flying Corps as a mechanic, he became a commissioned pilot in 1917, flying DH.5s and SE.5s until he was shot down over the Western Front. In the immediate post-war period he flew for the Avro Transport Company at Margate, before retiring from aviation to run a welding business in London until 1927. Drawn back into aviation, initially as chief instructor for the Henderson Flying School at Brooklands, with two other instructors, he bought the business in 1928, renaming it the Brooklands School of Flying. Over the following years, he grew the business into Brooklands Aviation Ltd, success coming with pleasure flying, pilot training and air shows at Brooklands which featured parachutists and 'wing-walkers'.

Dixey-Gerrans, Ronald

Born in Muswell Hill on 10th January 1903, Dixey-Gerrans first became known in aviation circles as secretary of the Herts and Essex Gliding Club in 1931. He then founded the joy-riding company, Aviation Developments Ltd in 1933 whereupon his aviation ambitions took a major step forward with the formation in 1934 of the Flying Fair Air Circus. Though less impressive than the major displays undertaken by Cobham and Scott, it nevertheless visited 150 towns in southern England in 1934.

Doig, Robert Galloway

'Dotty' Doig was a successful motor cycle racer before he took up commercial flying in Canada. After returning to Britain, he formed Air Limousines Ltd in 1934 to build a "clean-looking" aircraft, which seemingly never came to light. In 1935, he set up Universal Aircraft Company with allegedly a factory in Sidcup building Pou du Ciel (Flying Flea) aircraft and a 4-seater "fast" aircraft. His initial Flea had actually been built by EG Perman & Co but his second one, G-AEEW, had brief success with the British Empire Air Display before suffering a fatal crash at Penshurst on 4th May 1936. Universal became Aircraft Constructions Ltd in December 1936 and imported the Dutch Scheldemusch in 1937, which was also intended for licence-building. In 1938, he acquired the incomplete De Bruyne Lady Bird G-AFEG and renamed it the Doig Lady Bird, but this was no more successful. In 1939, he become a test pilot for the Peterborough Aircraft Co Ltd, which had bought up the assets of the defunct British Aeronca company, but this venture also soon went into receivership. In 1964, living in Kent, he claimed in a letter to Air Pictorial magazine that he was rebuilding a pre-war Corben Baby Ace and still had the Lady Bird project.

Eskell, Neil Dallas Louis

Although not a flying participant in the joy-riding arena, Dallas Eskell played a key managerial role in both Sir Alan Cobham's and CWA Scott's display tours of the United Kingdom. Born on 16th November 1880 in Bath, in the house once inhabited by the African adventurer, Dr Livingstone, he attended Bath College, played rugby for Bath, cricket with Somerset and served during the First World War with the Royal Flying Corps, being shot down twice over the Western Front. After post-war appointments with Instone Airlines and Imperial Airways, Eskell joined Cobham's organisation as general manager and later, served in a similar capacity with Charles Scott's touring display. Eskell's subsequent career embraced Olley Air Service, and after the Second World War, Field Aircraft Services and Morton Air Services at Croydon. In 1950, in a surprising change of career, he took on the management of the Aga Khan's racing stud in Ireland, but returned to the aviation fold in 1954 to join Silver City Airways. Eskell retired in 1965 and died on 3rd November 1973, just two weeks after Sir Alan Cobham.

Fielden, Earl Bateman

'Batty' Fielden was born on 14th October 1899 and became an instructor at the Yorkshire Aeroplane Club in 1926 before moving on to fly with Berkshire Aviation Tours Ltd in 1928 and Surrey Flying Services in 1929. In 1930, Fielden was the founder, managing director and chief pilot for joy-riding Aviation Tours Ltd which became a component part of National Aviation Day in 1932 but was dismissed after he was judged to have needlessly caused the crash of an Airspeed Ferry airliner. He joined the British Hospitals Air Pageant in 1933 and had agreed to charter Imperial Airways' old Handley Page W.8 airliner *"Prince Henry"*, but it was not made airworthy. In 1934, his successor company, Air Tours Ltd operated Imperial's Armstrong-Whitworth Argosy G-EBLF *"City of Glasgow"* for Barker & McEwan King's Sky Devils Air Circus. In 1935, he went on to undertake flying duties with United Airways Ltd and its successor British Airways.

Fisher, Francis Colebourne

'Bud' Fisher learned to fly with the London Aeroplane Club at Stag Lane in the mid-1920s and commenced his joy-riding career with Kent Aircraft Services in 1930. Soon afterwards he branched out on his own, operating his Avro 504K G-EBVL from Somerford Bridge, Christchurch (now in Dorset, but then in Hampshire) at what would later become the Airspeed Ltd factory site. Over a four season period (1930-1934) he carried some 19,000 passengers, two at a time in his Avro. In 1934, Fisher, along with Walter Burry, became more interested in establishing a proper airfield and Fisher Aviation Co was reformed as a limited company. In 1935, its assets were transferred to Bournemouth Airport Ltd when the airport was launched. Fisher's new commercial responsibilities as director of Bournemouth Airport's Flying School and Flying Club, left little time for joy-riding and G-EBVL, though having remained in a derelict condition on the site, was destroyed by the RAF in 1940. Fisher died sometime in either the late 1960s or early 1970s.

Fresson, Ernest Edmund

Ted Fresson was born on 20th September 1891 and his first joy-riding job was with the Cambridgeshire School of Flying at Skegness in 1919. Having earlier lived in China, he returned there in October 1919 and in the early 1920s was instrumental in putting up the first aircraft factory and testing the machines produced in the Chinese province of Shansi. Returning to England in 1927, he re-joined the joy-riding circuit with Berkshire Aviation Tours then based at Monksmoor, Shrewsbury. In 1929, Fresson, along with Lancelot Rimmer, co-founded North British Aviation Company Ltd at Hooton Park. Having realised the need for air-routes to be developed in Scotland, he formed Highland Airways Ltd in 1933, which ultimately became Scottish Airways Ltd in 1937. He continued to run the operation throughout the war and up until it was nationalised in 1947 to become part of British European Airways. A thorn in the flesh of the new management, he was sacked in 1948 but was allowed to retain his DH Gipsy Moth G-AAWO, with which he had founded Highland Airways. Fresson moved to Kenya in 1948 and died on 25th September 1963, shortly after writing his autobiography, 'Air Road to the Isles'.

Giroux, Sidney Norman

Norman Giroux, a French Canadian born in Quebec on 2nd December 1898, arrived in England in 1917 and flew with the Royal Flying Corps and Royal Air Force. Re-entering civilian life, he joined the International Aviation Company in Liverpool and served as a pilot for a year before, with his brother, Albert Percy Giroux, purchasing the locally-based Golden Eagle Aviation Co. Re-naming it the Giro Aviation Company he operated it for the next 19 years from Hesketh Park Aerodrome on the beach at Southport. A number of pilots who flew for the venture included John Nesbitt-Dufort, who, along with Giroux also flew with the RAF during the Second World War. Giroux re-established his business after the war, bringing his Fox Moths G-ACCB and G-ACEJ out of store, to finally retire in 1966 after a flying career that included some 30,000 flying hours and over 80,000 landings and take-offs. He died in 1993 aged 95.

Gower, Pauline Mary de Peauly

Born on 22nd July 1910, and the daughter of Sir Robert Gower MP, Pauline Gower learned to fly at the Phillips and Powis School of Flying at Woodley aerodrome, Reading in 1930. Having met Dorothy Spicer at Stag Lane, a formidable partnership was formed in which Miss Gower undertook the piloting and Miss Spicer the ground engineering duties. Joy-riding was carried out, firstly by the ladies on their own as Air Trips in Kent and later, while based at Hunstanton. During the early 1930s, Air Trips joined up with the larger air displays then touring the United Kingdom, including Modern Airways' Crimson Fleet (1932) and the British Hospitals Air Pageant (1933). Miss Gower carried well over 25,000 joy-riders over a five year period. Her proficiency as a pilot may be gauged from her appointment in 1936 as chief pilot in the British Empire Air Display team that included many of the country's leading display pilots. She became District Commissioner for the Eastern Division of the Civil Air Guard in May 1939 and the Head of the Women's Section of the Air Transport Auxiliary (ATA) in December 1939, where she remained until the unitwas disbanded in October 1945. Pauline Gower married Wing Cdr William Fahie in June 1945 and unfortunately died after giving birth to twins on 2nd March 1947.

Hearn, Martin Neito

Born in June 1906, Hearn began his career as an engineer but later gained national fame as the leading 'wing-walker' with Sir Alan Cobham's touring air displays. It was his fearless performances that probably inspired the less gifted imitators whose consequent mishaps resulted in 'wing-walking' being finally banned in mid-1933. Hearn was originally a pilot with the Lancashire School of Aviation Ltd and was later with Berkshire Aviation Tours prior to joining Sir Alan Cobham. In December 1937, he set up his own aeronautical engineering business, Martin Hearn Ltd, at Hooton Park, Cheshire and in 1938 became involved locally with Utility Airways Ltd, sometimes unkindly referred to as 'Futility Airways'! Owned by Frank Davison and his wife Joy (formerly Miss Joy Muntz, cousin of Airwork's Alan Muntz), Utility operated a varied fleet of aircraft including four Avro Cadets on pleasure flying at Blackpool in the late 1930s. In 1939, Martin Hearn Ltd gained its first contract for aircraft assembly work and performed major servicing and repairs on a wide variety of types throughout the war. Though becoming a director of Aerial Transit Ltd in 1946, Hearn became disenchanted with the government's bureaucratic direction of civil aviation and he left the industry to resume a successful business career elsewhere. Martin Hearn died on 2nd June 1992.

Henderson, George Lockhart Piercy

Henderson had an intensive ten-year career as a joy-riding pilot and flying instructor before his death in the crash of a Junkers F.13, G-AAZK belonging to Walcot Air Lines in July 1930. Following an early post-war spell as a pilot with the Avro Transport Co, in June 1919, he was put in overall charge of its nationwide joy-riding programme. In the late summer of 1919, he left to take a mixed fleet of machines to Sweden for joy-riding and mail carrying, returning to Britain a year later. After flying with Daimler Airways, he refused to join Imperial Airways on the merger of independent airlines in 1924, but did briefly become chairman of the pilots' strike committee which delayed the commencement of services. He set up his own business to instruct on Avro machines at Croydon until he moved to Brooklands to form the Henderson School of Flying in 1926. After he sold out in 1928, he became involved with the construction of light aircraft with designer John Bewsher and financier AP Glenny. In May 1930, he set up a sales and charter agency, Henderson Aviation Bureau, at Croydon, which after his death became British Air Transport Ltd under the management of his mother and his chief pilot Charles Allen.

Holmes, Frederick James Vernon (Fred)

Fred Holmes was born in June 1886 and his association with aviation commenced with an apprenticeship with AV Roe in 1912. He accompanied pioneer aviator FP Raynham as his mechanic during the *"Daily Mail"* seaplane tour of Great Britain in 1914. Immediately following this he joined the Royal Naval Air Service and remained on active service until his demobilisation in 1919 from what had then become the Royal Air Force. Fred Holmes founded Berkshire Aviation Co in 1919 with his brother Jack and stayed with the organisation until it merged with Northern Air Lines Ltd to form Northern Air Transport Ltd in April 1929. He later founded Air Travel Ltd in 1932 at Penshurst, moving to Gatwick in 1934. The company was taken over by Airports Ltd in 1936 but he stayed on as works and service manager at Gatwick. He died in April 1969.

Holmes, John Duncan Vernon (Jack)

Fred's younger brother, Jack Holmes, was born on 3rd June 1898 and learned to fly in the Royal Flying Corps. He fought with No.19 Squadron before being shot down in May 1917 and interned at Schweidniz, (Baron Manfred von Richthofen's home town) for the remainder of the war. Following his post-war association with Berkshire Aviation and later British Petroleum, he re-enlisted in the RAF at the outbreak of the Second World War. He attained the rank of Squadron Leader, became a Sector Operations Controller, serving at all the key stations in Fighter Command during and after the Battle of Britain. He died in 1980.

Ingham, Percy Hall

Percy Ingham received his flying instruction at the Northern Aircraft Co Ltd at Bowness-on-Windermere, but left without formal qualification to join the Royal Naval Air Service in 1916. Following a number of accidents he was invalided out of the service, never to (officially) pilot an aircraft again. It was the visit of Berkshire Aviation's 'Flying Festival' in February 1920 that persuaded Ingham to co-found Border Aviation with Bob and Graham Little, and to co-opt several Carlisle businessmen to invest in the new venture. However, Ingham's disconcerting attitude to business matters, (his wife reported that he threw unpaid bills into the fire) largely contributed to the demise of Border Aviation in October 1920. Whilst he was a party to the "phoenix" business, Ingham & Little Aviation Co formed immediately after, this too, only lasted for a single season.

Johnson, Hugh Coleman

'Johnnie' Johnson, born on 29th January 1907, served in the RAF as a Flight Lieutenant before becoming a demonstration pilot for the Ford Motor Company in 1931. After joining Cobham's National Aviation Day tour in 1932, he soon became chief pilot and went on to accompany the tours of South Africa and the UK from 1933-35. After the NAD disbanded, he became a founder member of Cobham's Flight Refuelling Ltd and played a leading part in the development of air-to-air refuelling in the late 1930s. Following wartime duties with the Royal Canadian Air Force, Johnson re-joined Flight Refuelling Limited as manager of flying administration and was placed in charge of the company's operations during the Berlin Airlift between July 1948 and August 1949. Johnson died on 28th December 1976.

Jones, Idwal

Little is recorded about the career of Idwal Jones except that, following a spell as a flying instructor at Brooklands, he became chief pilot for North Eastern Airways Ltd. Jones featured as aerobatic pilot on the touring air displays conducted by Sir Alan Cobham and C W A Scott in the early 1930s but died in the crash of Airspeed Courier G-ACSZ on 29th May 1937. In 1936, he built a Flying Flea, which remains to this day preserved at Caernarfon Air Museum.

Jones, Oscar Philip

Born in Beckenham, Kent on 15th October 1898, "OP" Jones learned to fly in the Royal Flying Corps. In January 1920 he joined the Berkshire Aviation Co Ltd, but the following year set up his own joy-riding operation with Fred Holmes, trading as Holmes and Jones Aviation Co. Jones left to join Instone Air Line in March 1922 and remained with the successors, Imperial Airways in 1924 and later BOAC in 1940 where he eventually became senior pilot. He died on 21st June 1980.

Kent, Frederick A

Freddie Kent was born on 8th January 1900 and though not a notable flier, brought considerable engineering expertise to the companies engaged in joy-riding. Inspired by Colonel Sam Cody, whose machine he helped to manhandle, at the age of 14 he became apprenticed to the Royal Aircraft Factory at Farnborough before entering the Royal Naval Air Service. In 1922, he joined Daimler Airways and became staff member No.2 in Imperial Airways when Daimler, along with Instone Airlines, Handley Page Air Transport and British Marine Air Navigation were merged in 1924. Kent left to join Surrey Flying Services as chief engineer in 1927, but rejoined Imperial Airways in 1930 to operate the Handley Page W8 for Aviation Tours. In 1933, Kent joined Captain 'Bill' Rollason as a partner in Rollason Aircraft Services. For what he described as 'an adventurous period', Kent served as chief engineer to Sir Alan Cobham's touring air display while still employed within the Rollason organisation. The outbreak of war saw him recruited as Scottish Aviation's production manager at Prestwick and in 1944 he managed the new BOAC repair division at Croydon. He retired in 1963 and died in 1996.

King, Joseph Richard

'Joe' King pursued a lively and colourful career both as a racing driver and a pilot. He flew under contract for the King of the Hedjaz [now effectively Saudi Arabia] and in Egypt and later for the Bolivian Government in the late 1920s. He then became an instructor at the Brooklands School of Flying in 1929 before joining National Flying Services in 1931. He participated in the Spartan Air Circus tour of South Africa in 1931-32 and became a regular touring pilot on the Cobham NAD display circuit, flying the Airspeed Ferry. He later transferred to CWA Scott's 'Flying For All' as chief pilot in 1936.

Kingwill, Archibald Norman

Captain Kingwill was born on 5th September 1898 and learned to fly in the Royal Flying Corps in 1917. After leaving the Service he co-founded Aerial Photos Ltd in 1919 and joined the band of 'barnstorming' pilots that mushroomed in the post-war period, forming the Kingwill and Jones Flying Company in 1921. Following this short-lived venture he became chief pilot for a five-year period (1923-28) with William Beardmore & Co Ltd at Renfrew. Whilst there, he co-founded the Scottish Flying Club in 1928 before moving south of the border to join, as chief pilot, Fred Holmes' Manchester based joy-riding organisation which in 1932 became Air Travel Ltd. Kingwill retained his interest in aerial photography, founding Aero Pictorial in June 1934 and went on to become Sales Manager for Airports Ltd in 1937. During the 1930s, he was a test pilot for several short-lived private concerns, including CLW Aircraft with their Curlew and, in 1939, the Willoughby Delta Company which produced an experimental twin-boom, two-seater aircraft.

Lawson, Harold

Born in Newcastle, Lawson served with 4 (Naval) Squadron, RNAS before eventually becoming an instructor at the RAF's Central Flying School. Returning to civilian life, he managed timber works in China and Lagos, Nigeria before joining Berkshire Aviation Tours Ltd in 1928. Spells of employment followed with Northern Air Lines and Aviation Tours Ltd, during which he accompanied Sir Alan Cobham's display tours in Britain in 1932 and South Africa 1932-33. It was in the closing stages of the visit to South Africa that, on 17th February, Lawson took his own life and that of his passenger by diving his Avro Tutor G-ABZR into the centre of Cape Town's Wynberg airport.

Lowe-Wylde, Charles Herbert

Born in Newcastle in 1901, 'Jimmy' Lowe-Wylde soon demonstrated a practical interest in aeronautics, building his first glider at the age of fourteen. He later became apprenticed to Sir WG Armstrong Whitworth and, after a period spent in the RAF, worked successively for Blackburns, Supermarine and Short Brothers. Attracted by the growing glider movement in the late 1920s, Lowe-Wylde and four companions within the space of five weeks in 1930 created the first all-British designed glider. Encouraged by this, he founded The British Aircraft Company in Maidstone, Kent, to design and manufacture a series of gliders and which, in 1932 became BAC Ltd. Lowe-Wylde pioneered the launching of gliders by towing them behind motor cars, demonstrations of which proved very popular at Cobham's National Aviation Day displays. In 1932 he developed a motorised version of his BAC.VII glider by mounting a Douglas 600cc motorcycle engine above the wing and fitted with a rudimentary undercarriage, this became known as the Planette. It was while demonstrating Planette BGA186 at West Malling aerodrome on 13th May 1933 that Lowe-Wylde crashed and was killed. He was thirty-two years old. The inquest report recorded that his real name was Thomas Harold Lowe and he assumed the name Lowe-Wylde when he commenced his own business operations in 1926.

McEwan King, James

Jimmy McEwan King played a leading role in the formation of various joy-riding concerns in the early 1930s. Originally a printer by trade with the family firm of Galbraith King & Co Ltd, in 1932 he co-founded Air Pageants Ltd in partnership with Harry Barker. They traded as British Hospitals Air Pageants in 1933 and under various names in each year thereafter – Sky Devils Air Circus (1934), Jubilee Air Displays (1935), British Empire Air Displays (1936) and in 1937 as The Coronation Air Display. The last named venture proved a financial failure, but McEwan King was involved in a number of other aviation ventures, including Anderson Aerocars Ltd, which became Deekay Aircraft Corporation in 1936. During the war, he was a founding director of King Aircraft Corporation, which then manufactured aircraft seats.

McIntosh, Robert Henry

Captain McIntosh, familiarly referred to as 'All-Weather Mac' was born in 1894 and served in the Royal Flying Corps in 1917 and subsequently the RAF. In September 1919, he joined Handley Page Transport and its successor, Imperial Airways in 1924. In 1927, he took part in unsuccessful attempts by Commandant James Fitzmaurice to cross the Atlantic east to west and later with Bert Hinkler to fly non-stop to India - on both occasions using Fokker F.VIIa G-EBTS. After appointments as private pilot to Major Andrew Holt of Aircraft Investment Corporation and The Hon Mrs Florrie Westenra, he again teamed up in 1932 with Bert Hinkler to tour Canada and the USA in a Lockheed Vega. In 1933, he accompanied the British Hospitals Air Pageant, flying a DH.60M Moth and went on to fly with Spartan Air Lines and its successors, United Airways and British Airways. He died in 1983.

Mackay, William

'Jock' Mackay was born in 1901 and became a partner with Lance Rimmer in North British Aviation Co at Hooton Park in 1929. He joined the Cobham tour in 1933 and 1934 but died from injuries suffered as a result of the crash of Avro 504N G-ACRS near Farnborough on 30th June 1934.

Muir, Arthur Frederick

Following wartime service, Captain Muir became an instructor and joy-riding pilot with Leatherhead Aviation Services in 1920 before branching out to become co-owner of Surrey Flying Services in 1921. Muir joined 'Bill' Rollason in 1932 to form Rollason Muir and Rickard Aviation Company of Croydon and this later became Rollason Aviation Company Ltd and participated in the Cobham tours. He resigned from Rollasons in 1933 but came back into aviation when he was appointed an instructor to the Nottingham Flying Club in 1939.

Nash, Thomas William James

Tommy Nash joined Rollason Aviation Company in 1931, but resigned to form the Barnstaple and North Devon Flying Club at Heanton Court in December 1933. He later rejoined the Rollason organisation to become, in 1935, pilot and sales manager for Rollason Aircraft Services Ltd. The following year, he founded Nash Aircraft Sales and, in 1937, Anglo-European Airways Ltd, a Croydon-based charter operator. Always a popular figure, he piloted a Rollason-owned Desoutter Mk I on the Cobham and Scott touring displays.

Parkinson, JD

After service with the Royal Flying Corps and the Royal Air Force, Captain 'Parky' Parkinson began his joy-riding career with Berkshire Aviation Company Ltd in 1924 and then became a flying instructor with the Newcastle-upon-Tyne Aero Club from July 1926 to September 1928. After emigrating to Canada, he joined International Airways, and later Canadian Airways before becoming a test pilot with Curtiss-Reid in

Montreal. Returning to England in 1932, he formed part of the Cornwall Aviation Company's detachment that accompanied Cobham's NAD tour of the UK. Parkinson later entered flying administration, becoming the movements control officer at Perth in Scotland in 1936 before taking up similar appointments at Croydon, Portsmouth, Southampton Water, Heston and Bristol (Whitchurch). He died on 6th March 1943.

Phillips, Percival

Born in July 1893, Cornishman Phillips initially took up training as an automobile engineer before volunteering for the Duke of Cornwall's Light Infantry in 1914. The following year he transferred to the Royal Flying Corps as a mechanic, but he later became an observer and, after further training, a pilot, serving in Mesopotamia, Persia and Kurdistan. Upon return to civilian life in November 1919, Captain Phillips became a partner in the Hill and Phillips Garage in St Austell. Having undertaken the overhaul of Berkshire Aviation Company's engines, Phillips decided to re-enter aviation and in May 1924 formed the Cornwall Aviation Co with one Avro 504K and Albert Adams as ground engineer. The company enjoyed considerable success and expanded throughout the 1920s when many others failed to survive, though this was largely due to a regular income and engineering resources being supplied from Phillips' garage. In 1932, Phillips, aware of the publicity attending the forthcoming tour by Sir Alan Cobham, agreed to guarantee the appearance at each venue of two Cornwall Aviation Avros. In practice, this meant supplying three machines to cover maintenance 'down-time' periods. After a similar spell of contract work in 1933 with the British Hospitals Air Pageant, Phillips returned to accompany Cobham's team in 1934. In 1935 the formation of Air Publicity Ltd based at Heston, took up much of Phillips' time, but he formed a partnership with CWA Scott to purchase most of Cobham's aircraft and equipment at the end of the year. This formed the basis for 'CWA Scott's Flying Display Ltd' which, though impressive, enjoyed only limited financial success in 1936. Air Publicity did however secure a large share of the towed banner advertising market throughout the 1930s, but the staid nature of the flying did not appeal to Phillips' inherently adventurous spirit. Sadly, it was this exuberant streak that seemingly contributed to his death in a flying accident at Gamlingay on 13th February 1938.

Pixton, Cecil Howard

Born on 14th December 1885 at Didsbury, Manchester, Pixton was apprenticed at the Industrial Engineering Company at Newton Heath. In 1910, he joined AV Roe & Co and learned to fly as an unpaid mechanic. Quickly promoted to chief instructor at the company's Brooklands school, he moved in 1911 to The British & Colonial Aeroplane Company. In 1914, he gained international acclaim by winning the Schneider Trophy at Monte Carlo in a Sopwith Tabloid seaplane. Immediately after the War, Pixton ran the Avro Transport Company's flying service at Windermere with joy-riding and passenger carrying in the North West area and Wales, combining with similar activities on the Isle of Man. He took this operation over as the short-lived Lakes Motor & Seaplane Company in 1920 but was, in 1928, on the ground staff of Lancashire Aero Club and in the 1930s served in a similar role at Haldon in Devon. He died on 7th February 1972.

Rimmer, Lancelot John

Born on 19th February 1896, Lance Rimmer, a Canadian who arrived in the United Kingdom with the Canadian Expeditionary Force during the First World War, was destined to become a leading figure in the inter-war flying fraternity. Having gained his wings in 1918, he became a pilot with International Aviation Co on the Isle of Man in 1920. Between 1926 and 1928, he was a pilot with Berkshire Aviation Tours Ltd, although he also was involved in the grandiosely-named joy-riding business, South Wales Airways, in 1927. In 1929, he co-founded another joy-riding business, North British Aviation Co Ltd at Hooton. In 1933, with Jock Mackay, he purchased seven 504Ks from the liquidation of Northern Air Transport and flew several of them with Cobham. In 1936, he joined CWA Scott and in 1937, Coronation Air Displays. In 1939, he set up Ace Air Services at Speke to operate a Monospar ST.4 on charters and became a test pilot on Halifaxes with Rootes during World War II.

Rollason, William Arthur

'Bill' Rollason was born on 22nd June 1899 and by 1930 was chief pilot at the Dudley Watt School of Flying. In October 1930, he set up the first of a succession of companies bearing his name that engaged in flying instruction, air taxi work and joy-riding. He, along with colleagues Freddie Kent and Tommy Nash were closely associated with the British Hospitals Air Pageant and Cobham's National Aviation Day tours. In 1935, he also became technical director with Crilly Airways. He set up WA Rollason Ltd, at Croydon, in May 1940, a company forever associated after the war with converting ex-RAF Tiger Moths for civilian use.

Scott, Charles William Anderson

Charles Scott, born on 13th February 1903, learned to fly during his four-year spell in the RAF (1922-26). After leaving the Service, he went to Australia, where he flew with QANTAS until 1930 when he returned to the UK. In 1931 he flew solo to and from Australia in a DH Gipsy Moth and in 1932 carried out another flight to Australia followed by a tour of Russia. Scott's fame rests largely on the success he achieved with Tom Campbell Black in winning the MacRobertson Air Race from England to Australia in October 1934, for which they became joint winners of the Royal Aero Club's Britannia Trophy. Capitalising on his celebrity, Scott made many guest appearances on the touring air display circuit before founding, with Captain Percy Phillips, CWA Scott's Flying Display Ltd in December 1935. In addition to participating in the King's Cup and Schlesinger Air Races in 1936, his other ventures included becoming aviation editor of the News Chronicle and founding an aeroplane ferry company, Aircraft Collection and Delivery Ltd., in 1939. Scott committed suicide in Berlin on 15th April 1946.

Spicer, Dorothy Norman

Born on 31st July 1908, Dorothy Spicer, became a pilot and, more particularly, a fully qualified ground engineer. Irrevocably associated with Miss Pauline Gower in the early 1930s, they jointly founded Air Trips for joy-riding in 1932 and enjoyed success on UK tours with Modern Airways' Crimson Fleet and the British Hospitals Air Pageant. Dorothy Spicer's unique engineering skills – she was then the only woman in the world to hold the Air Ministry's 'A', 'B', 'C' and 'D' Licences – authorised her to inspect, pass out and repair both engines and airframes. She was in fact qualified to build an aircraft or engine from scratch and to approve all materials necessary for the work! Her talent complemented the flying skills of Miss Gower and enabled them to undertake their own joy-riding operations at Hunstanton, Norfolk in 1934 and 1935. Miss Spicer became an inspector for the Air Registration Board in 1938 and during the Second World War she worked at the Royal Aircraft Establishment as an Air Observer and Research Assistant. Having left England to start a new post-war life in Brazil, she died along with her husband, Richard Pearse, when the FAMA Avro York LV-XIG crashed into a mountainside on 23rd December 1946.

Summerfield, Samuel

Sammy Summerfield was operating pleasure flight operations throughout the 1920s, initially under his own name as S Summerfield & Co on the East Coast. In 1927, he joined Cornwall Aviation Company, flying Avro 504K G-EBSE before leaving at the end of 1929 to set up Wolverhampton Aviation Co Ltd.

Taylor, Joseph Carey Crabtree

'Joe' Taylor was born on 20th February 1896 and became a joy-riding pilot with the AV Roe Company at Southport in 1919 before joining the Berkshire Aviation Company in 1920. In 1924 he took part as co-pilot in the Oxford University Arctic Expedition and in 1926, became aviation manager for the British Petroleum Company Ltd, where he had an extensive career

both before and after the Second World War. During that conflict, however, Taylor commanded the Elementary Gliding School at Denham.

Tranum, John

Born in Denmark in 1900, John Tranum *(left, below)* worked as a stuntman in America before arriving in England in 1930. He then joined the British Russell Parachute Company to demonstrate the Russell Lobe parachute, but moved to the Irving Air Chute Company a year later. He then co-founded Skywork Ltd. and went with the Spartan Air Circus tour of South Africa. Later the same year, he was the main parachutist with CD Barnard's 'Barnardstormers'. He achieved the free-fall record on 24th May 1933 from 17,250 feet but died after suffering a heart attack at altitude just prior to attempting another record free-fall parachute jump near Copenhagen, Denmark on 7th March 1935.

Turner Hughes, Charles Keir

'Toc-H' as he was more familiarly known, was born on 27th June 1904 and became chief pilot for Caribbean Airways based in Jamaica in 1931. He returned to Britain and joined Cobham's NAD tour of the UK in 1932 and the winter tour of South Africa in 1932-33. His skill as chief pilot was unrivalled until Geoffrey Tyson succeeded him in 1933. After leaving Cobham's employment, Turner Hughes entered industry as a test pilot with Armstrong-Whitworth Aircraft Ltd., and served as chief test pilot from 1936-46.

Tyson, Geoffrey Arthur Virney

Born in 1907, Tyson spent his early years in Purley, attending the Whitgift School and perhaps, more significantly, becoming inclined towards a flying career while watching the activity at nearby Croydon Airport. After a brief period spent articled to an estate architect, Tyson learned to fly at 5 Flying Training School, Sealand and served with the RAF between 1926 to 1931. Following this he became a flying instructor, firstly with the Maidstone Aero Club at West Malling and later at the Scarborough Aero Club. In the early 1930s he joined Sir Alan Cobham's National Aviation Day tours and achieved much public acclaim with his low-level inverted flying and general aerobatic prowess. After a spell as an experimental pilot with AV Roe in 1935-37, he furthered his association with Cobham by becoming involved with early air-to-air refuelling development. During the Second World War, Tyson carried out production flight testing of Sunderland flying boats and Stirling heavy bombers while working for Short Brothers. Post-war he was appointed chief test pilot for Saunders Roe based on the Isle of Wight and became famously associated with test flying the giant Princess flying boat and SR.A/1 flying boat jet fighter. He died in January 1987.

Ward, Harry

Though born in London in 1903, Harry Ward was brought up in Yorkshire and joined the RAF in 1921. By 1926 he had become part of the team demonstrating parachute techniques at stations around the country. Leaving the Service in 1929, Ward joined the London General Omnibus Company and whilst there, helped form its own flying club. Ward went on to become a principal parachutist with all the major touring displays throughout the 1930s. He re-enlisted in the RAF when war became imminent and became part of the unit formed by Colonel Louis Strange to train airborne forces at Ringway's Parachute School. Ward re-entered civilian life in 1945 but never failed to fly whenever the opportunity arose. He died on 24th July 2000.

Warren-Merriam, Frederick

Born in 1880 as Frederick George Warren, he changed his name by deed poll in 1901. Having first been engaged in the motor business, he became a flying instructor at the Bristol Flying School in 1911. After wartime service in the Royal Naval Air Service, he joined the Avro Transport Company at Hamble in April 1919, flying both the 504K and the float-equipped 504L at resorts on the Isle of Wight and Hayling Island. In 1921, Warren-Merriam became a pilot for SE Saunders Ltd., test flying the Kittiwake flying boat. The following year he, along with Saunders' works manager, G Newman built the Merriam-Newman glider for entry in the *"Daily Mail"* Itford Competition. He also founded the Whitely Bank School of Gliding at Cowes in 1923 before moving on in 1926 to become an aviation consultant, trading as Merriam's Aviation Bureau. In 1928, he contemplated starting up a seaplane and landplane school, but this did not materialise although he provided joyrides from the West Cowes aerodrome. He died in 1956.

Typical of the early joy-riding concerns, here is Berkshire Aviation Company visiting Wisbech in the mid-1920s.

APPENDIX IV
The Touring Itineraries

This Appendix lists the known touring itineraries of the major organisations that operated across the United Kingdom. The recorded details still available vary considerably and in certain cases the listings are far from complete.

Dates marked * are confirmed days within a longer visiting period when the full dates are not known

Venues are listed, where known, in the order visited, though in many instances the precise dates cannot now be ascertained.

Dates	Location

BERKSHIRE AVIATION COMPANY [1919-1922]

1919

Dates	Location
27-29 April	Thatcham, NEWBURY
-	WANTAGE
-	THAME
-	Race Meadow, AYLESBURY
-	Whitley Home Farm, Upper Redlands Road, READING
11 [?] June	Haremore Park, FARINGDON
-	ABINGDON
-	SWINDON
-	Harper's Field, Rose Hill, OXFORD
-	LEIGHTON BUZZARD
-	Jackman's Farm, BEDFORD
30-31 July	Tickler's Midsummer Meadow, NORTHAMPTON
1-6 Aug	Aylestone Lane, LEICESTER
7-10 Aug	Big Meadow, Meadow Lane, LOUGHBOROUGH
11-17 Aug	Sneinton Junction Meadow, Trent Lane, NOTTINGHAM
25-30 Aug	Hell Meadow, Sunny Hill, DERBY
-	STOKE-ON-TRENT
-	Ashgate, nr VAD Hospital, CHESTERFIELD
-	Sumerleyton Avenue, KIDDERMINSTER
-	Monkmoor Aerodrome, SHREWSBURY
9-15 Oct	Bridgnorth Road, Compton, WOLVERHAMPTON
-	Calderfields Farm, Mellish Rd, Longwood Canal Bridge, WALSALL
1-5 Nov	Polo Ground, Barby Road, RUGBY
-	LEAMINGTON
17-23 Nov	Priestthorpe Lane Aerodrome, Leeds Rd, Farsley, BRADFORD
24-30 Nov	Newham Bridge Farm, Linthorpe, MIDDLESBROUGH
1-21 Dec	Town Moor Aerodrome, NEWCASTLE-UPON-TYNE
22 Dec*	Crew Toll, Ferry Road, EDINBURGH

1920

Dates	Location
26 Jan-4 Feb	Dalston Road, Cummersdale Turn, CARLISLE
20 Feb-1 Mar	Rating Lane, Furness Abbey, BARROW-IN-FURNESS
1-14 March	Ribble Bridge, Walton-le-Dale, PRESTON
-	BOLTON
1-14 April	Chester Road, WARRINGTON
-	MACCLESFIELD
10 Jul-14 Aug	Lock's Common, PORTHCAWL
15-30 Aug	Greenway Road, TAUNTON
1-6 Sept	FROME
7-23 Sept	BANBURY
24 Sept*	Castle Meads, GLOUCESTER
11-25 Oct	Billersley Aerodrome, Kings Heath, BIRMINGHAM
26 Oct-9 Nov	Port Hill Meadow, STOKE-ON-TRENT
6-16 Dec	Swallowfield Acre, Vicarage Road, NORTHWICH

1921

Dates	Location
1-14 Jan	Pennyhouse Lane, Burnley Road, ACCRINGTON
-	Nr White Horse Inn, Leeds Road, HUDDERSFIELD
-	Coal Aston Aerodrome, Jordanthorpe, SHEFFIELD
23 Feb-6 March	Tiln Lane, RETFORD
7-17 March	Canal Bridge, CASTLEFORD
18-31 March	NEWARK
1-13 April	Ashby Road, Hugglescote, COALVILLE
14-24 April	The Headlands, KETTERING
25 Apr-6 May	South Hill Farm, Mill Road, WELLINGBOROUGH
-	HITCHIN
-	ST ALBANS
-	DUNSTABLE
-	HERTFORD
1 July -	Newton Road, RUSHDEN [Northants]
13 July*	Sayers Marsh, KINGS LYNN
23 Jul-6 Aug	HULL [Robinson/Taylor only – Holmes in S Wales]
1-8 Sept	Hood's Field, Branstone, BURTON-ON-TRENT
10-24 Sept	Hay Road, HEREFORD
28 Sept-7 Oct	GLOUCESTER
-	Castle Bromwich Aerodrome, BIRMINGHAM
-	ASHBOURNE
-	RUGELEY
-	SHIFNAL
-	NEWPORT
-	WELLINGTON

1922

Dates	Location
1-13 Jan	WEM
14-22 Jan	Ash Road, WHITCHURCH
5-12 Feb*	NANTWICH

-	Sych Farm Meadow, MARKET DRAYTON
-	CONGLETON
-	Easters Meadow, LEOMINSTER
-	Ludford House Meadow, LUDLOW
-	TEWKESBURY
10 May*	Islington Farm, WELLS [Somerset]
-	TIVERTON
15 June*	NEWTON ABBOT
16 June	TAVISTOCK
-	Braunton Road, BARNSTAPLE
-	Racecourse at Laira Bridge, PLYMOUTH
-	BIDEFORD
-	SHERBORNE
-	BLANDFORD FORUM
-	CHEPSTOW
5-12 Sept	Actis Meadow, GLASTONBURY [Tor Fair Week]
27 Sept-14 Oct	NEWENT [Gloucester Barton Fair]
-	LONGTON
-	UTTOXETER
11 Nov*	LICHFIELD
26 Nov*	Tamworth Road, SUTTON COLDFIELD
-	BASINGSTOKE
-	Waterloo Crossing, WOKINGHAM
20-26 Dec	Betchworth Park, DORKING
27 Dec	REDHILL

SIR ALAN COBHAM'S MUNICIPAL AERODROME CAMPAIGN
Tour of the UK – May to October 1929
The following list details the locations in the order visited:

OXFORD
BEDFORD
SPALDING
BOSTON
Mousehold Aerodrome, NORWICH [19th May]
NORTHAMPTON
Woodley Aerodrome, READING [22nd-23rd May]
STRATFORD-ON-AVON
Filton Aerodrome, BRISTOL
BATH
HUNTINGDON
SHEFFIELD
RIPON
HARROGATE
YORK
LINCOLN
BEVERLEY
NEWCASTLE
CAMBRIDGE
WORCESTER
WOLVERHAMPTON
WARRINGTON
CREWE
Aberkinsey Farm, Dyserth Road, RHYL [27th June]
LEEDS
HULL
BRIDLINGTON
STOCKTON-ON-TEES
GLASGOW
STIRLING
PERTH
EDINBURGH
LANARK
DUMFRIES
STRANRAER
AYR
SUNDERLAND
SCARBOROUGH
SKEGNESS
BURTON-ON-TRENT
DONCASTER
NOTTINGHAM
SOUTHPORT
LANCASTER
Walney, BARROW-IN-FURNESS [1st August]
ISLE OF MAN
BLACKPOOL
South Denes, GREAT YARMOUTH
MAIDSTONE
Swalecliffe Aerodrome, WHITSTABLE [11th August]
Swingate Aerodrome, DOVER [13th August]
Baddow Meads, CHELMSFORD [14th August]
CANTERBURY
TUNBRIDGE WELLS
EASTBOURNE
BRIGHTON

ISLE OF WIGHT
SALISBURY
YEOVIL
BOURNEMOUTH
WEYMOUTH
DORCHESTER
Haldon Aerodrome, TEIGNMOUTH [27th August]
Polo Ground, Roborough, PLYMOUTH
Northern Burrows, BIDEFORD
NEWPORT
CARDIFF
CIRENCESTER
FOLKESTONE
Hartsdown, MARGATE [7th-8th September]
BOGNOR REGIS
TEWKESBURY
LUDLOW
Lower Lugg Flats, HEREFORD [12th September]
BASINGSTOKE
CHATHAM
AYLESBURY
COLCHESTER
WOLVERTON
ST ALBANS
FELIXSTOWE
LUTON
WORTHING
PORTSMOUTH
Brockworth Aerodrome, GLOUCESTER
WINDSOR
MAIDENHEAD
LOUTH
GRIMSBY
LOWESTOFT
HITCHIN
Cofton Hackett Landing Ground, BIRMINGHAM
DUNSTABLE
SOUTH SHIELDS
MIDDLESBROUGH
LETCHWORTH
STEVENAGE
MORECAMBE
CHICHESTER
STONY STRATFORD
NEWPORT PAGNELL
HERNE BAY
TONBRIDGE
LITTLEHAMPTON
GILLINGHAM
ROCHESTER
CHELTENHAM

AVIATION TOURS LTD
Tour of Britain – 1931
Only incomplete itinerary known

4-13 April	Port Meadow Aerodrome, Wolvercote, OXFORD
14-23 April	Billings Meadow, Great Haughton, NORTHAMPTON
24 Apr-3 May	Braunston Aerodrome, LEICESTER
-	SHEFFIELD
-	The Racecourse, DONCASTER
-	The Knavesmire Racecourse, YORK
18-27 May	Morton Palms, Yarm Road, MIDDLESBROUGH
-	NEWCASTLE
31 May	Whitchurch Airport, BRISTOL [Opening of Airport]
-	EDINBURGH
-	GLASGOW
-	INVERNESS
August	SOUTHPORT SANDS
4 August	Hedon Aerodrome, HULL [for return of Amy Johnson]
5 August	SOUTHPORT SANDS
-	PENZANCE
late Sept	Tour ended

NORTH BRITISH AVIATION CO LTD
Tour of Britain – 1931
[Incomplete itinerary known]

27 Apr-3 May	Ely Racecourse, CARDIFF
10 May	Whitley Aerodrome, COVENTRY
20-21 May	Gretna Green
5-7 June	Moorpark Aerodrome, Renfrew, GLASGOW
14 June	Broxbourne Aerodrome, HODDESDON
29 July	Wash Farm, CLACTON
9 August	DOVERCOURT
26 August	The Aerodrome, CLACTON

5 September	The Aerodrome, Gorleston, GREAT YARMOUTH
13 September	Crabb's Farm, Hamstel Road, Southchurch, SOUTHEND
17 September	Broxbourne Aerodrome, HODDESDON
20 September	Crabb's Farm, Hamstel Road, Southchurch, SOUTHEND
27 September	Crabb's Farm, Hamstel Road, Southchurch, SOUTHEND

CD BARNARD AIR TOURS LTD
Tour of the UK – 1931

1 April	LUTON
2-3 April	SOUTHAMPTON
4-5 April	COVENTRY
6-7 April	SOUTHPORT
11-12 April	MAIDSTONE
18-19 April	Mousehold Aerodrome, NORWICH
25-26 April	Ratcliffe Aerodrome, LEICESTER
27 April	HITCHIN
29 April	Woodley Aerodrome, READING
-	STRATFORD-ON-AVON
-	NEWARK
-	WOMBWELL
-	BARNSLEY
-	YORK
-	LEEDS
-	BEVERLEY
-	SPALDING
-	CAMBRIDGE
-	BEDFORD
-	OXFORD
13-14 May	Chapel Farm, Lansdown, BATH
-	FROME
-	PLYMOUTH
-	YEOVIL
-	STOKE
-	WALLINGTON
-	BLACKPOOL
29 May	Barton Aerodrome, MANCHESTER
1 June	BROUGH
-	DONCASTER
-	RIPON
-	GRANTHAM
-	CASTLE BROMWICH
-	LEICESTER
-	NOTTINGHAM
-	IPSWICH
-	HINCKLEY
-	STOCKTON-ON-TEES
-	FLEETWOOD
-	NEWCASTLE
-	ISLE OF MAN
-	Whitchurch Aerodrome, BRISTOL
-	GLOUCESTER
-	ILFRACOMBE
-	CAMELFORD
-	BIDEFORD
-	REDRUTH
-	WESTON-SUPER-MARE
27-28 June	Splott Foreshore Aerodrome, CARDIFF
-	DEVIZES
30 June ?	ASHFORD
16 July	Church Farm, Stoney Hills, Heacham, HUNSTANTCN
23 July	DORCHESTER
-	Atlantic Park Aerodrome, Eastleigh, SOUTHAMPTON
4-5 August	CHICKERELL
14 August	Chapel Farm, Lansdown, BATH
15-17 August	Heston Air Park, LONDON
3 September	WIGTON
21 September	STRANRAER
22 September	DUMFRIES
25 September	KENDAL
11 October	HANWORTH AIR PARK

MODERN AIRWAYS LTD/THE CRIMSON FLEET
Tour of the UK – 1932
Few details are recorded about this tour

14 May	Shoreham Aerodrome, BRIGHTON
16 May	Beech Farm, Kipping Cross, Pembury, TUNBRIDGE WELLS
-	BATH
[1 week]	Westland Aerodrome, YEOVIL
-	WHITCHURCH

3-11 June	Splott Foreshore Aerodrome, CARDIFF
-	Castle Bromwich Aerodrome, BIRMINGHAM
-	Speke Airport, LIVERPOOL
15 June	Stanley Park Aerodrome, BLACKPOOL
August	PORTHCAWL
30 August	Haldon Airfield, TEIGNMOUTH
3 October	-

SIR ALAN COBHAM'S NATIONAL AVIATION DAY/ DISPLAY TOURS 1932-1935
1932

April 12	Hanworth Aerodrome, LONDON [Tour assembly point]
13 April	LUTON
14 April	Morden Grange, ROYSTON
15 April	Cardington Airfield, BEDFORD
16 April	Goodmayes Park, ILFORD
17 April	Hanworth Aerodrome, LONDON
18 April	Old Golf Course, BASINGSTOKE
19 April	High Post Aerodrome, SALISBURY
20 April	Woodley Aerodrome, READING
21 April	Stroud Farm, Bray, MAIDENHEAD
22 April	AYLESBURY
23-24 April	Low Halls Farm, WALTHAMSTOW
25 April	Airfield, LEIGHTON BUZZARD
26 April	Warwick Road Airfield, BANBURY
27 April	Hoo Farm, KIDDERMINSTER
28 April	Perdiswell Park, WORCESTER
29 April	Black Horse Barn, Farringdon Road, ABINGDON
30 April-1 May	Stag Lane Aerodrome, EDGWARE
2 May	SWINDON
3 May	TROWBRIDGE
4 May	WELLS
5 May	FROME
6 May	CIRENCESTER
7-8 May	Parton Farm, Churchdown, GLOUCESTER
9 May	Caulderfields Farm, Mellish Rd, WALSALL
10 May	Chestnut Farm, 42 Acre, Sandy Lane, MELTON MOWBRAY
11 May	Stenson Road Flying Ground, DERBY
12 May	Tollerton Aerodrome, NOTTINGHAM
13 May	Rushley Farm, Derby Rd, MANSFIELD
14 May	Lowford Heath, RUGBY
15-16 May	Old Polo Field, Bramley's Farm, Cockfosters Rd, ENFIELD
17 May	Maylands Aerodrome, ROMFORD
18 May	Broomfield Aerodrome, CHELMSFORD
19 May	Blue Barns Aerodrome, COLCHESTER
20 May	Biggin Hill Aerodrome, WESTERHAM
21-22 May	West Malling Aerodrome, Kingshill, MAIDSTONE
23 May	Bekesbourne Aerodrome, CANTERBURY
24 May	St Margarets Aerodrome, Swingate Downs, DOVER
25 May	Star Farm, CHATHAM
26 May	Goodmayes Park, ILFORD
27 May	Whincups Field, Marlow Hill, HIGH WYCOMBE
28-29 May	Northfield Aerodrome, Cofton Hackett, BIRMINGHAM
30 May	Oldfield off Kings Acre Road, HEREFORD
31 May	Rock Parc Dol, LLANDINDROD WELLS
1-2 June	CARDIFF
3 June	Flying Ground, The Front, PORTHCAWL
4 June	Whitchurch Aerodrome, BRISTOL
5-6 June	Chapel Farm, Lansdown, BATH
7 June	Wernllwyd Farm, Newton Rd, WELSHPOOL
8 June	Borras Lodge Farm, Bieston, WREXHAM
9 June	Meir Aerodrome, STOKE-ON-TRENT
10 June	Barton Aerodrome, MANCHESTER
11 June	Speke Aerodrome, LIVERPOOL
12 June	Cop House Farm, East Saltney, CHESTER
13 June	Stafford Common, STAFFORD
14 June	Spittlegate Hill, Saltersford, GRANTHAM
15 June	Boardsides, Sleaford Road, BOSTON
16 June	Armthorpe Aerodrome, DONCASTER
17 June	St John's Heath, Bracebridge, LINCOLN
18 June	Yeadon Aerodrome, BRADFORD
19 June	Sherburn-in-Elmet Aerodrome, LEEDS
20 June	Middleton Park, LEEDS
21 June	Waltonwrays, Carleton Fields, SKIPTON
22 June	The Racecourse, RIPON
23 June	Mete House Farm, Walton Bridge, Fishwick, PRESTON
24-25 June	Scale Hall, Morecambe Rd, LANCASTER
26-27 June	Stanley Park Aerodrome, BLACKPOOL
28 June	The Foreshore, Fleetwood, WHALLEY
29 June	Agricultural Show Field, KENDAL
30 June	Orton Grange, Wigton Road, CARLISLE
1 July	Greenfield Moor Farm, Morpeth Rd, ALNWICK
2-3 July	Cramlington Aerodrome, NEWCASTLE

4 July	WEST HARTLEPOOL
5 July	Flying Field, Cargo Fleet Lane, MIDDLESBROUGH
6 July	East Leys Farm, Grindale, BRIDLINGTON
7 July	Old Racecourse, SCARBOROUGH
8-9 July	Coal Aston Aerodrome, Jordanthorpe, SHEFFIELD
10 July	Old Wombwell Aerodrome, BARNSLEY
11 July	Old Racecourse, GOOLE
12 July	Royal Oak Airfield, Roman Bank, SKEGNESS
13 July	The Flying Ground, KETTERING
14 July	Desford Airfield, LEICESTER
15 July	Flying Field, Cubbington Road, LEAMINGTON SPA
16 July	Fountain Hotel Aerodrome, Loughton, BLETCHLEY
17 July	Whitley Abbey Aerodrome, COVENTRY
18 July	Cote Hill, Husbands Bosworth, MARKET HARBOROUGH
19 July	Westwood Aircraft Park, Castor Hill, PETERBOROUGH
20 July	Church Farm, Stoney Hills, Heacham, HUNSTANTON
21 July	Lodge Farm, Croxton Road, THETFORD
22 July	Laurel Farm, North Repps, CROMER
23 July	Mousehold Aerodrome, NORWICH
24 July	Wheatcroft Farm, Gorleston, GREAT YARMOUTH
25 July	Municipal Aerodrome, Nacton Road, IPSWICH
26 July	The Flying Ground, Alton Park Road, Plough Corner, CLACTON
27 July	Alderstead Farm, MERSTHAM
28 July	Heston Airport, LONDON
29 July	Parsonage Farm, Swalecliffe, HERNE BAY
30 July	Coldblow Farm, Walmer Road, DEAL
31 Jul-1 Aug	Nethercourt Flying Ground, London Rd, RAMSGATE
2-3 August	Frowd's Flying Field, Kings Drive, EASTBOURNE
4 August	North Heath Farm, HORSHAM
5 August	Rookery Farm, Lower Kingswood, REIGATE
7-8 August	Shoreham Aerodrome, BRIGHTON
8 August	Ford Aerodrome, Yapton, LITTLEHAMPTON
9 August	Chalcraft Farm, North Bersted, BOGNOR
10 August	Municipal Aerodrome, PORTSMOUTH
11 August	Atlantic Park Aerodrome, Eastleigh, SOUTHAMPTON
12-13 August	Iford Bridge, Castle Lane, BOURNEMOUTH
14 August	Chickerell Aerodrome, WEYMOUTH
15 August	Haldon Aerodrome, TEIGNMOUTH
16 August	Three Holes Cross, Camelford Rd, WADEBRIDGE
17 August	Davidstow Tylands Corner, CAMELFORD
18 August	Home Farm, Tehidy Park, CAMBORNE
19 August	Great Rosevidney Farm, Long Rock, Cockwells, PENZANCE
20 August	Roborough Aerodrome, PLYMOUTH
21 August	Follygate Airfield, OKEHAMPTON
22 August	Whalesborough, BUDE
23 August	West Stowfold Farm, Westdown, ILFRACOMBE
24 August	Musgrove Farm, Wellington Rd, TAUNTON
25 August	Woodspring Priory, Sand Bay, WESTON-SUPER-MARE
26 August	Pershore Racecourse, EVESHAM
27 August	Racecourse Farm, Llanfoist, ABERGAVENNY
28 August	Vennaway Lane, Fairwood Common, Penard, SWANSEA
29 August	HAVERFORDWEST
30 August	Morfa Mawr, ABERAYRON
31 August	Harlescott Flying Ground, SHREWSBURY
1 September	Tynewedd Farm, Capel Curig Rd, BANGOR
2 September	Aberkinsey Farm, Dyserth Rd, RHYL
3 September	Merrills Bridge, CREWE
4 September	Flying Ground, Chester Rd, WARRINGTON
5 September	Squires Gate Aerodrome, ST ANNES
6 September	Low Houses Farm, WIGTOWN
7 September	Tinwald Downs, DUMFRIES
8 September	Teviot Bridge, KELSO
9 September	Sherriff Muir, Easter Happrew, PEEBLES
10-11 Sept	Silverknowes, Davidson's Mains, EDINBURGH
12 September	Westbank Farm, LANARK
13 September	The Old Aerodrome, ALLOA
14 September	Banbeath, Windygates-Leven Road, Buckhaven, STIRLING
15 September	Rennyhill, ANSTRUTHER
16 September	Lennoxlove Airfield, Acredales, Gifford Rd, HADDINGTON
17-18 Sept	Moorpark Aerodrome, Renfrew, GLASGOW
19 September	West Mains Farm, Grangemouth Rd, FALKIRK
20 September	Balgrove, Cupar Road, ST ANDREWS
21 September	Woodhead of Mailer, PERTH
22 September	Seafield, Longman Road, INVERNESS
23 September	Mid Ardlaw, New Petsligo Road, FRASERBURGH
24 September	East Seaton Links, ABERDEEN
25-26 Sept	The Barns of Claverhouse, Forfar Road, DUNDEE

27 September	The Aerodrome, MONTROSE
28 September	Elliot Brae, Netherkelly, Forfar-Dundee Rd, ARBROATH
29 September	Heatherstacks, Brechin Road, FORFAR
30 September	Methil, Kirkcaldy-Leven Road, BUCKHAVEN
1 October	Broomhouse Farm, Corstorphine, EDINBURGH
2 October	Scremstone Hill Farm, Newcastle Rd, BERWICK
3 October	Holderhouse Farm, Cleadon, SUNDERLAND
4 October	Bennet Lane Fields, Leeds Road, DEWSBURY
5 October	Oak View, Harrogate-Wetherby Rd, Plompton, HARROGATE
6 October	Hedon Aerodrome, HULL
7 October	BOLSOVER
8 October	Coal Aston Aerodrome, Jordanthorpe, SHEFFIELD
9 October	BIRMINGHAM
10 October	Irchester Grange, WELLINGBOROUGH
11 October	CAMBRIDGE
12 October	The Aerodrome, GRAVESEND
13 October	DARTFORD
14 October	SHEERNESS
15 October	WOOLWICH
16 October	CHINGFORD

Tour of the Union of South Africa 1932-1933

23-26 Nov	CAPE TOWN
27-30 Nov	Not Known
1-4 December	CAPE TOWN
5 December	MOOREESBURG
6 December	WELLINGTON
7 December	WORCESTER
8 December	ROBERTSON
9 December	RIVERSDALE
10-11 Dec	OUDTSHOORN
12 December	WILLOWMORE
13 December	GRAAF REINET
14 December	MIDDLEBURG
15 December	CRADOCK
16 December	TARKASTAD
17-18 Dec	QUEENSTOWN
19 December	ALIWAL NORTH
20 December	BETHULIE
21 December	COLESBURG
22 December	FAURIESMITH
23-24 Dec	BLOEMFONTEIN
25 December	none
26-27 Dec	KIMBERLEY
28 December	LADYBRAND
29 December	FICKSBURG-BETHLEHEM
30 December	VEREENIGING
31 Dec-1 Jan	BARAGWANATH

1933

2 January	BENONI
3 January	RANDFONTEIN
4 January	WITBANK
5 January	PRETORIA
6 January	NYLSTROOM
7-8 January	GERMISTON
9 January	RUSTENBURG
10 January	VENTERSDORP
11 January	KLERKSDORP
12 January	KROONSTAD
13 January	PARYS
14 January	POTCHEFSTROOM
15 January	none
16 January	HEILBRON
17 January	FRANKFURT
18 January	BETHAL
19 January	ERMELO
20 January	STANDERTON
21 January	HARRISMITH
22 January	none
23 January	NEWCASTLE
24 January	DUNDEE
25 January	LADYSMITH
26-27 Jan	PIETERMARITZBURG
28-29 Jan	DURBAN
30 January	none
31 January	KOKSTAD
1-3 February	none
4 February	EAST LONDON
5 February	none
6 February	GRAHAMSTOWN
7 February	SOMERSET EAST
8 February	none

9 February	UITENHAGE
10 February	none
11 February	GEORGE
12 February	none
13 February	MOSSEL BAY
14 February	SWELLENDAM
15 February	BREDASDORP
16 February	WELLINGTON
17 February	CAPE TOWN

1933 – No.1 Tour

14-15 April	Central Park, DAGENHAM
16-17 April	Shooters Hill, WOOLWICH
18 April	Ickleford, HITCHIN
19 April	Gatwick Aerodrome, HORLEY
20 April	Bekesbourne Aerodrome, CANTERBURY
21 April	TENTERDEN
22-23 April	Alderstead Farm, Shepherds Hill, MERSTHAM
24 April	WALLINGFORD
25 April	Chalvey, SLOUGH
26 April	WELWYN
27 April	PANGBOURNE
28 April	Headley Court, EPSOM DOWNS
29-30 April	Hamsey Green Aerodrome, WARLINGHAM
1 May	FARINGDON
2 May	Witney Aerodrome, OXFORD
3 May	UXBRIDGE
4 May	COLCHESTER
5 May	The Vale, Cheltenham Road, EVESHAM
6-7 May	Cofton Hackett Aerodrome, BIRMINGHAM
8 May	Newnham Grounds, DAVENTRY
9 May	BANBURY
10 May	KIDDERMINSTER
11 May	BURTON-ON-TRENT
12 May	The Old Aerodrome, Nottingham Rd, MELTON MOWBRAY
13-14 May	Bishops Tachbrook Aerodrome, LEAMINGTON
15 May	Handley Park Farm, TOWCESTER
16 May	RUGBY
17 May	Whincups Field, Marlow Hill, HIGH WYCOMBE
18 May	ASHFORD [Kent]
19 May	Sincox Lane, THAKEHAM
20 May	Little Common, BEXHILL
21 May	FOLKESTONE
22 May	LINGFIELD
23 May	Palmers Farm, Shenfield, BRENTWOOD
24 May	ST ALBANS
25 May	OXFORD
26 May	FINEDON
27 May	IPSWICH
28 May	CLACTON
29 May	Standford's Farm, Queenborough Lane, BRAINTREE
30 May	Old Racecourse, Portholme Meadow, HUNTINGDON
31 May	BOSTON
1 June	Aerodrome, CAMBRIDGE
2 June	LICHFIELD
3 June	Coal Aston Airfield, Jordanthorpe, SHEFFIELD
4 June	NOTTINGHAM
5 June	Coal Aston Airfield, Jordanthorpe, SHEFFIELD
6 June	NEWARK
7 June	Waddington Aerodrome, LINCOLN
8 June	Armthorpe Airfield, DONCASTER
9 June	BAWTRY
10 June	The Aerodrome, WOODFORD
11-12 June	Middleton Park, LEEDS
13 June	WREXHAM
14 June	Cop House Farm, CHESTER
15-16 June	Fern Hill Farm, Gobowen, RHOS UCHA
17-18 June	BIRKENHEAD
19 June	KNUTSFORD
20 June	BLACKPOOL
21 June	LANCASTER
22 June	KENDAL
23 June	WHITEHAVEN
24-25 June	WORKINGTON
26 June	WIGTOWN
27 June	ANNAN
28 June	CASTLE DOUGLAS
29 June	STRANRAER
30 June	Travelling to Ireland
1-2 July	Kildonan Airfield, Finglas, DUBLIN
3 July	WATERFORD
4 July	CLONMEL
5-6 July	Ballincollig, CORK
7-8 July	Raheen, Ballycummin Castle, LIMERICK
9 July	DUBLIN

10 July	Oranmore Airfield, GALWAY
11 July	BUNDORAN
12-13 July	LONDONDERRY
14-15 July	Sydenham, BELFAST
16 July	DUNDALK
17 July	Return from Ireland
18 July	HAWICK
19 July	LOCKERBIE
20 July	CARLISLE
21 July	PENRITH
22 July	Mountbatten, BARROW-IN-FURNESS
23 July	Aintree Racecourse, LIVERPOOL
24 July	Harehill Farm, HEYWOOD
25 July	Siddon's Farm, Ainsworth Rd, BURY
26 July	Back of Moss Farm, UNSWORTH
27 July	STOKE-ON-TRENT
28 July	Manor House Farm, Timperley, ALTRINCHAM
29 July	REDDITCH
30 July	Kitchen Lane, Wednesfield, WOLVERHAMPTON
31 July	NUNEATON
1 August	LOUGHBOROUGH
2 August	GREAT BARR
3 August	BURTON-ON-TRENT
4 August	STRATFORD-ON-AVON
5-7 August	Stag Lane Aerodrome, EDGWARE
8 August	HERNE BAY
9 August	MARGATE
10-11 August	FOLKESTONE & DOVER
12 August	Upper Stoneham Farm, South Malling, LEWES
13 August	BRIGHTON
14 August	Wilmington Aerodrome, EASTBOURNE
15 August	Ford Aerodrome, Yapton, LITTLEHAMPTON
16 August	BURGESS HILL
17 August	RYDE
18 August	Apse Aerodrome, SHANKLIN
19 August	PORTSMOUTH
20 August	Canford Village, Magna Road, WIMBORNE
21 August	PETWORTH
22 August	East Parley, BOURNEMOUTH
23 August	WEYMOUTH
24 August	SWANAGE
25 August	Haldon Aerodrome, TEIGNMOUTH
26 August	PLYMOUTH
27-28 August	Big Field, Trebelzue, NEWQUAY
29 August	Home Farm, Tehidy Park, CAMBORNE
30 August	Rosevidney Farm, Cockworth, PENZANCE
31 August	Racecourse Farm, Six Chimneys, BODMIN
1 September	Between Whalesborough Farm & the Coast, BUDE
2 September	North Devon Aerodrome, BARNSTAPLE
3 September	Park Farm, West Buckland, WELLINGTON
4-5 Sept	Mullacott, Moortown, ILFRACOMBE
6 September	Rexworthy Farm, Durleigh Road, BRIDGEWATER
7 September	NAILSWORTH
8 September	HAY-ON-WYE
9 September	Gregory's Farm, Ash Hall, WHITCHURCH [Salop]
10 September	BEAUMARIS
11 September	Travelling to Ireland
12 September	Ballybar, CARLOW
13 September	Killiane, WEXFORD
14 September	Ballinamuck, DUNGARVAN
15 September	The Aerodrome, FERMOY
16 September	Cardenton, ATHY
17 September	Portmarnock, DUBLIN
18 September	Ballymore, BOYLE
19 September	Ellesmere Park, BUNDORAN
20 September	Scardenmore, SLIGO
21 September	Old Aerodrome, CASTLEBAR
22 September	The Racecourse, BALLINROSE
23 September	Rathnaleen, NENAGH
24 September	Western Park, KILLARNEY
25 September	Ballinorig, TRALEE
26 September	Dongeeha, NEWCASTLE WEST
27 September	The Racecourse, MALLOW
28 September	Kilcommon, CAHIR
29 September	Kinnity Road, TULLAMORE
30 September	Big Meadow, ATHLONE
1 October	Colpe Farm, DROGHEDA
2 October	Return from Ireland
3 October	Griffiths Crossing, CAERNARVON
4 October	PRESTEIGN
5 October	Perdiswell Park Aerodrome, WORCESTER
6 October	Rock Hill Farm?, CHIPPING NORTON
7 October	Upper Farm, MOLESEY
8 October	Chertsey Lane, STAINES

1933 – No.2 Tour

Date	Location
14-15 April	Holt Farm, Ashingdon Rd, SOUTHEND
16 April	The Aerodrome, GRAVESEND
17 April	West Malling Aerodrome, MAIDSTONE
18 April	Aldenham Country Club, Ham Farm, ELSTREE
19 April	Lewsey Farm, Dunstable, LUTON
20 April	HERTFORD
21 April	KINGS LANGLEY
22-23 April	Sussex Farm?, West Clandon, GUILDFORD
24 April	ALTON
25 April	PETERSFIELD
26 April	Rookery Farm, LOWER KINGSWOOD
27 April	CHICHESTER
28 April	NEWBURY
29 April	Atlantic Park Aerodrome, Eastleigh, SOUTHAMPTON
30 April	Burry's Farm, CHRISTCHURCH
1 May	LYMINGTON
2 May	Cann Common, SHAFTESBURY
3 May	SALISBURY
4 May	Staverton Aerodrome, GLOUCESTER
5 May	WARMINSTER
6-7 May	TAUNTON
8 May	Heanton Court, Chivenor Farm, BARNSTAPLE
9 May	AA Landing Ground NGR4243, DELABOLE
10 May	NEWLYN EAST
11 May	Field on the A30, HAYLE
12 May	Ventonwyn Farm, ST AUSTELL
13 May	EXETER
14 May	RAF Folly Gate, OKEHAMPTON
15 May	WELLS
16 May	SHERBORNE
17 May	DEVIZES
18 May	RADSTOCK
19 May	WESTON-SUPER-MARE
20-21 May	Oldfield, off Kings Acre Road, HEREFORD
22 May	MELKSHAM
23 May	CHEPSTOW
24 May	PORT TALBOT
25 May	CARDIGAN
26 May	Rest Day
27 May	PORTHCAWL
28 May	CARDIFF
29 May	BUILTH WELLS
30 May	ABERAYRON
31 May	WELSHPOOL
1 June	Caulderfields, WALSALL
2 June	LEOMINSTER
3-4 June	BRISTOL
5 June	Stenson Fields, DERBY
6 June	STONE
7 June	MALVERN
8 June	Sundorne Farm, SHREWSBURY
9 June	BRIDGENORTH
10-11 June	SOLIHULL
12 June	TAMWORTH
13 June	ST NEOTS
14 June	Leverington Common, WISBECH
15 June	Maysland, Great Easton, GREAT DUNMOW
16 June	NEWPORT PAGNELL
17-18 June	BRIGHTON
19 June	SUDBURY
20 June	LONG EATON
21 June	SLEAFORD
22 June	RETFORD
23 June	SCUNTHORPE
24 June	Hedon Aerodrome, HULL
25 June	Brough Aerodrome, HULL
26 June	Barton Aerodrome, MANCHESTER
27 June	LEEK
28 June	REDCAR
29 June	WEST HARTLEPOOL
30 June	CHESTER-LE-STREET
1-2 July	Cramlington Aerodrome, NEWCASTLE-ON-TYNE
3 July	KELSO
4 July	PENICUIK
5 July	PERTH
6 July	KIRKINTILLOCH
7 July	Old Aerodrome, ALLOA
8-9 July	GLASGOW
10 July	BATHGATE
11 July	ST ANDREWS
12 July	BRECHIN
13 July	KIRRIEMUIR
14 July	HUNTLY
15 July	Bloodymires, MACDUFF
16-17 July	East Seaton Links, ABERDEEN
18 July	LOSSIEMOUTH
19 July	NAIRN
20 July	STONEHAVEN
21 July	BATHGATE
22-23 July	EDINBURGH
24 July	DUNBAR
25 July	NORTH BERWICK
26 July	AIRDRIE
27 July	BERWICK-ON-TWEED
28 July	MORPETH
29-30 July	TYNEMOUTH
31 July	SUNDERLAND
1 August	RICHMOND [Yorks]
2 August	Morton Palms Farm, DARLINGTON
3 August	STOCKTON-ON-TEES
4 August	East Leys Farm, Grindale, BRIDLINGTON
5 August	Yeadon Aerodrome, LEEDS
6-7 August	BLACKBURN
8 August	BLACKPOOL
9 August	BOLTON
10 August	THIRSK
11 August	YORK
12-13 August	BIRKENHEAD
14 August	BAKEWELL
15 August	Winthorpe Aerodrome, SKEGNESS
16-17 August	Church Farm, Aylmerton, CROMER
18 August	Corton Road, LOWESTOFT
19 August	NORWICH
20 August	GREAT YARMOUTH
21 August	FRINTON
22 August	Holt Farm, Ashingdon Rd, SOUTHEND
23 August	Nethercourt Flying Ground, RAMSGATE
24 August	BIRCHINGTON
25 August	BEXHILL
25 August	Wilmington Aerodrome, EASTBOURNE
27 August	Chalcraft Farm, North Berstead, BOGNOR REGIS
28 August	BRIDPORT
29 August	EXETER
30 August	BUDLEIGH SALTERTON & EXMOUTH
31 August	LYME REGIS
1 September	Pennygillan, LAUNCESTON
2 September	Porthmissen Farm, PADSTOW
3 September	Lankelly Farm, FOWEY
4 September	CHARD
5 September	MINEHEAD
6 September	CHELTENHAM
7 September	AMMANFORD
8-9 September	TENBY
10 September	ABERGAVENNY
11 September	PWLLHELI
12 September	Rhydorddwy Fawr, RHYL
13 September	WHALLEY
14 September	PRESTON
15 September	WIDNES
16 September	Coal Aston Aerodrome, Jordanthorpe, SHEFFIELD
17 September	BARNSLEY
18 September	Eastburn Farm, DRIFFIELD
19 September	MANSFIELD
20 September	Waltham Aerodrome, GRIMSBY
22 September	Westwood Aircraft Park, PETERBOROUGH
23 September	EAST DEREHAM
24 September	Court Yard Farm, Ringstead, HEACHAM
25 September	NORTH WALSHAM
26 September	Captain Wilson's Field, Little Downham Rd, ELY
27 September	Mettingham?, BUNGAY
28 September	SAFFRON WALDEN
29 September	BLETCHLEY
30 September	Hook Aerodrome, KINGSTON-UPON-THAMES
1 October	HARROW
2 October	Five Ash Down?, UCKFIELD
3 October	EAST GRINSTEAD
4 October	CHIPPENHAM
5 October	ALDBOURN
6 October	HORSHAM
7 October	DARTFORD
8 October	Maylands Aerodrome, ROMFORD

1934 Tour

Date	Location
14 April	Central Park, DAGENHAM
15 April	Stoke Hill Estate, West Clandon, GUILDFORD
16 April	ALTON
17 April	DEVIZES
18 April	WESTON-SUPER-MARE
19 April	BATH

Date	Location
20 April	MALVERN
21 April	LEAMINGTON SPA
22 April	BIRMINGHAM
23 April	LEEK
24 April	LYMM
25 April	BOLTON
26 April	COCKERMOUTH
27 April	WHITEHAVEN
28 April	CARLISLE
29 April	Mountbatten, BARROW-IN-FURNESS
30 April	PENRITH
1 May	DUMFRIES
2 May	KELSO
3 May	LANARK
4 May	KIRKINTILLOCH
5 May	CAMBUSLANG
6 May	Davidsons Mains, EDINBURGH
7-8 May	DUNDEE
9 May	MACDUFF
10 May	PETERHEAD
11 May	HUNTLY
12-13 May	East Seaton Links, ABERDEEN
14 May	INVERNESS
15 May	THURSO
16 May	WICK
17 May	TAIN
18 May	LOSSIEMOUTH
19 May	BERWICK-ON-TWEED
20-21 May	TYNEMOUTH
22 May	CHESTER-LE-STREET
23 May	Skerningham Farm, Harrogate Hill, DARLINGTON
24 May	Coal Aston Aerodrome, Jordanthorpe, SHEFFIELD
25 May	KNOTTINGLEY
26 May	Armthorpe Airfield, DONCASTER
27 May	SUNDERLAND
28 May	The Racecourse, SCARBOROUGH
29 May	LOUTH
30 May	RETFORD
31 May	SCUNTHORPE
1 June	MARKET WEIGHTON
2-3 June	Thornhills Lane, BRIGHOUSE
4 June	REDCAR
5 June	RICHMOND [Yorks]
6 June	DERBY
7 June	STOKE-ON-TRENT
8 June	WALSALL
9-10 June	BIRMINGHAM
11 June	BEDFORD
12 June	Whincups Field, Marlow Hill, HIGH WYCOMBE
13 June	Woodley Aerodrome, READING
14 June	CHESHAM
15 June	SHERBORNE
16 June	Whitchurch Airport, BRISTOL
17 June	BIRMINGHAM
18 June	RUGBY
19 June	Perdiswell Park Aerodrome, WORCESTER
20 June	KIDDERMINSTER
21 June	NOTTINGHAM
22 June	Hesketh Park Aerodrome, SOUTHPORT
23 June	The Aerodrome, WOODFORD
24 June	LEAMINGTON SPA
25 June	Ramsey, HARWICH
26 June	Blue Barns Aerodrome, COLCHESTER
27 June	Penshurst Aerodrome, TONBRIDGE
28 June	Airfield, SWALECLIFFE
29 June	PETERSFIELD
30 June	Cove, FARNBOROUGH [Hants]
1 July	The Aerodrome, GRAVESEND
2 July	Five Ash Down?, UCKFIELD
3 July	Sincock Lane, THAKEHAM
4 July	Atlantic Park Aerodrome, Eastleigh, SOUTHAMPTON
5 July	HAVANT
6 July	WHITECHURCH [Hants]
7 July	Wick Lane, Southborne, BOURNEMOUTH
8 July	Aerodrome, RYDE
9 July	Apse Aerodrome, SHANKLIN
10 July	Magna Road, Canford, WIMBORNE
11 July	CF Davis' Cemetery Farm, Salisbury Road, BLANDFORD
12 July	NEWBURY
13 July	TROWBRIDGE
14 July	Hanham, BRISTOL
15 July	Kitchen Lane, Wednesfield, WOLVERHAMPTON
16-17 July	ABERYSTWYTH
18 July	PWLLHELI
19 July	none
20 July	BIRKENHEAD
21 July	LIVERPOOL
22 July	Aintree Racecourse, LIVERPOOL
23 July	BLACKPOOL
24 July	BURY
25 July	BOLTON
26 July	BLACKBURN
27 July	UNSWORTH
28-29 July	MANCHESTER
30 July	Aerodrome, PERSHORE
1 August	Aerodrome, BARNSTAPLE
2 August	Big Field, Trebelzue, NEWQUAY
3 August	Rosevidney Farm, Cockwells, PENZANCE
4 August	MULLION
5 August	Portmission Farm, PADSTOW
6 August	Boskenso Farm, FALMOUTH
7 August	Sticker Aerodrome, Rocky Park, ST AUSTELL
8 August	EXETER
9 August	EXMOUTH
10 August	LYME REGIS
11 August	Haldon Aerodrome, TEIGNMOUTH
12 August	TAUNTON
13 August	CREWKERNE
14 August	SWANAGE
15 August	LITTLEHAMPTON
16 August	WINCHESTER
17 August	Chalcraft Farm, North Bersted, BOGNOR REGIS
18-19 August	Shoreham Aerodrome, BRIGHTON
21-22 August	HASTINGS
22 August	Eridge Road, TUNBRIDGE WELLS
23 August	Little Common, BEXHILL
24 August	Kingsnorth?, FOLKESTONE
25 August	The Aerodrome, ROCHESTER
26 August	BIRCHINGTON
27 August	ISLE OF SHEPPEY
28 August	Holt Farm, Ashingdon Rd, Rochford, SOUTHEND
29 August	FRINTON
30 August	Wheatcroft Farm Aerodrome, Gorleston, GREAT YARMOUTH
31 August	CROMER
1 September	Church Farm, Stoney Hills, Ringstead Rd, Heacham, HUNSTANTON
2 September	WAKEFIELD
3 September	Hedon Aerodrome, HULL
4 September	SALTBURN
5-6 Sept	STOCKTON-ON-TEES
7 September	WEST HARTLEPOOL
8 September	Cramlington Aerodrome, NEWCASTLE-UPON-TYNE
9 September	GATESHEAD
10 September	Middleton Park, LEEDS
11 September	Capt Wilson's Field, Little Downham Rd, ELY
12-13 Sept	SUTTON [Cambs]
14 September	NORTHAMPTON
15-16 Sept	TUNSTALL
17 September	LINCOLN
18 September	Mablethorpe, SUTTON-ON-SEA
19 September	BURTON-ON-TRENT
20 September	LEICESTER
21 September	Wellingborough Road, KETTERING
22 September	COVENTRY
23 September	Chalvey, SLOUGH
24 September	HEREFORD
25 September	CARDIFF
26 September	BARRY
27 September	PORT TALBOT
28 September	CHEPSTOW
29 September	STAINES
30 September	Maylands Aerodrome, ROMFORD

1935 Tour

Date	Location
12 April	Titchfield Road, FAREHAM
13 April	Kingsmill Aerodrome, REDHILL
14 April	Imperial College Sports Ground, East Lane, HARROW
15 April	Whitchurch Road, ANDOVER
16 April	Witney Aerodrome, OXFORD
17 April	AA Landing Ground, Gloucester Road, CIRENCESTER
18 April	Blackmore Park, MALVERN
19-20 April	Cottage Farm, Marston Green, BIRMINGHAM
21-22 April	Mile Flat, Greens Forge, DUDLEY
23 April	Astmore, RUNCORN
24 April	Langtree Hall Farm, Standish, WIGAN
25 April	Top o' the Moor Farm, Guide, BLACKBURN
26 April	Rose Farm, Wincham, NORTHWICH

Date	Location
27-28 April	Travers Farm, Bold, ST HELENS
29 April	Silvester Farm, German Lane, Euxton, CHORLEY
30 April	Smethwick Green, SANDBACH
1 May	Hooton Park Aerodrome, BIRKENHEAD
2 May	Bodvel Hall, PWLLHELI
3 May	Ballybar Aerodrome, CARLOW
4 May	Hollyfort, ENNISCORTHY
5 May	Marsh Farm, DUNDALK
6-7 May	BELFAST
8 May	COLERAINE
9 May	LONDONDERRY
10 May	Scardenmore, SLIGO
11 May	Leopardstown Racecourse, DUBLIN
12 May	Phoenix Park, DUBLIN
13 May	Bloomfield, MARYBOROUGH
14 May	Drinagh [Kenniscarthy?], WEXFORD
15 May	Dunmore, KILKENNY
16 May	Kilcohan Park Racecourse, WATERFORD
17 May	Ballycummin Castle, LIMERICK
18-19 May	Farmers Cross Aerodrome, CORK
20 May	The Racecourse, TRALEE
21 May	Oranmore Airfield, GALWAY
22 May	Old Aerodrome, CASTLEBAR
23 May	Big Meadow [Roscommon?], ATHLONE
24 May	Clooncoose Racecourse, LONGFORD
25 May	LIVERPOOL
26 May	BIRKENHEAD
27 May	BLACKPOOL
28 May	none
29 May	LEEDS
30 May	The Aerodrome, WOODFORD
31 May	RETFORD
1-2 June	Clifton, Thornhills Lane, BRIGHOUSE
3 June	Armthorpe Airfield, DONCASTER
4 June	COVENTRY
5 June	Stenson Fields, DERBY
6 June	WOLVERHAMPTON
7 June	TEWKESBURY
8 June	LEAMINGTON SPA
9-10 June	BIRMINGHAM
11 June	WESTON-SUPER-MARE
12 June	YEOVIL
13 June	BATH
14 June	Bray?, MAIDENHEAD
15 June	SOUTHAMPTON
16 June	ROCHESTER
17 June	THAME
18 June	OXFORD
19 June	Cockpole Green, HENLEY-ON-THAMES
20 June	Fen Ditton Aerodrome, CAMBRIDGE
21 June	CHESHAM
22 June	IPSWICH
23 June	Maylands, ROMFORD
24 June	HAVANT
25 June	FORDINGBRIDGE
26 June	SALISBURY
27 June	Rookery Farm, Lower Kingswood, REIGATE
28 June	ALTON
29 June	none
30 June	The Aerodrome, GRAVESEND

Display then split into No.1 and No.2 Tours:

No.1 Tour

Date	Location
1 July	Penshurst Aerodrome, TONBRIDGE
2 July	SITTINGBOURNE
3 July	MAIDSTONE
4 July	Five Ash Down?, UCKFIELD
5 July	Sincox Lane, THAKEHAM
6 July	Wick Lane, Southbourne, BOURNEMOUTH
7 July	Wilmington Aerodrome, EASTBOURNE
8 July	MALDON
9 July	WALTON-ON-TH-NAZE
10 July	FELIXSTOWE
11 July	Boggis's Farm, Lowestoft Rd, SOUTHWOLD
12 July	CROMER
13 July	NORWICH
14 July	Wheatcroft Farm Aerodrome, Gorleston, GREAT YARMOUTH
15 July	Maisdyke Lane, Fleet Church End, HOLBEACH
16 July	Sleaford Rd, Bourne, MORTON
17 July	SWADLINCOTE
18 July	SUTTON-ON-SEA
19 July	SCUNTHORPE
20 July	Newmillerdam, WAKEFIELD
21 July	Clifton, BRIGHOUSE
22 July	Hedon Aerodrome, HULL
23 July	WHITBY
24 July	SALTBURN
25 July	GATESHEAD
26-27 July	Woolsington Aerodrome, NEWCASTLE-UPON-TYNE
28 July	NOTTINGHAM
29 July	Church Farm, Stoney Hills, HEACHAM
30 July	Ramsey, HARWICH
31 July	Marlow Hill, HIGH WYCOMBE
1 August	Nethercourt Flying Ground, London Rd, RAMSGATE
2 August	BIRCHINGTON
3 August	Swalecliffe, HERNE BAY
4 August	FOLKESTONE
5-6 August	HASTINGS
7 August	BEXHILL
8 August	WORTHING
9 August	Chalcraft Farm, North Bersted, BOGNOR REGIS
10 August	Burry's Farm, Somerford Rd, CHRISTCHURCH
11 August	COWES
12 August	YARMOUTH
13 August	Lea Aerodrome, SANDOWN
14 August	The Aerodrome, BEMBRIDGE
15 August	SWANAGE
16 August	Chickerell Aerodrome, WEYMOUTH
17 August	Haldon Aerodrome, TEIGNMOUTH
18 August	Denbury Aerodrome, NEWTON ABBOT
19 August	LYME REGIS
20 August	EXMOUTH
21 August	TAVISTOCK
22 August	Pennygillan, LAUNCESTON
23 August	Lodge, [3 miles from] LISKEARD
24 August	Roborough, PLYMOUTH
25 August	Trebulzue Farm, NEWQUAY
26 August	Great Rosevidney, Cockwells, PENZANCE
27 August	Belle Vue, TORRINGTON
28 August	Sea Lane Meadow, Dunster, MINEHEAD
29 August	Lynmoor, LYNTON
30 August	Marsh Hill, DULVERTON
31 August	Magna Road, Canford, WIMBORNE
1 September	Lee-on-Solent Foreshore, GOSPORT
2 September	Church Norton, SELSEY
3 September	Wick Farm, LITTLEHAMPTON
4 September	Hook Aerodrome, KINGSTON-UPON-THAMES
5 September	Chertsey Lane, STAINES
6 September	Cosford, SHIFNAL
7-8 Sept	Warring's Field, Langtree Hall Farm, Standish, WIGAN
9 September	SOUTHPORT
10 September	Lowton, MANCHESTER
11 September	Aintree Racecourse, LIVERPOOL
12 September	Coal Aston, Jordanthorpe, SHEFFIELD
13 September	COALVILLE
14-15 Sept	SOLIHULL
16 September	Caldecote, Weddington Lane, NUNEATON
17 September	Handley Park Farm, Abthorpe Rd, TOWCESTER [cancelled]
18 September	Hoo Farm, KIDDERMINSTER
19 September	Shobnall Rd, BURTON-ON-TRENT
20 September	The Aerodrome, WALSALL
21 September	Sywell Aerodrome, NORTHAMPTON
22 September	Perton Court, WOLVERHAMPTON
23 September	Lawford Heath, RUGBY
24 September	Trading Estate, SLOUGH
25 September	Stonecott Hill, London Rd, SUTTON [Surrey]
26 September	Hainault Recreation Ground, ILFORD
27 September	Winkfield Plain Farm, WINDSOR
28 September	Whitehall Farm, Cove, FARNBOROUGH [Hants]
29 September	Phoenice Farm, Bagdon Hill, DORKING

No.2 Tour

Date	Location
1 July	Old Racecourse, Portholme Meadow, HUNTINGDON
2 July	STAMFORD
3 July	SELBY
4 July	NORTHALLERTON
5 July	CHESTER-LE-STREET
6-7 July	DURHAM
8 July	DUMFRIES
9 July	CASTLE DOUGLAS
10 July	CAMBUSLANG
11 July	HAMILTON
12 July	Kincardine Farm, CRIEFF
13-14 July	KIRKALDY
15 July	INVERNESS
16 July	DINGWALL
17 July	WICK
18 July	THURSO

Date	Location
19 July	LOSSIEMOUTH
20 July	MACDUFF
21 July	KINTORE
22 July	ST ANDREWS
23 July	KILMARNOCK
24 July	EDINBURGH
25 July	PEEBLES
26 July	KELSO
27 July	Woolsington Aerodrome, NEWCASTLE-ON-TYNE
28 July	SUNDERLAND
29-30 July	CARLISLE
31 July	WIGTOWN
1 August	PENRITH
2 August	WORKINGTON
3 August	Mountbatten, BARROW-IN-FURNESS
4 August	MIDDLETON
5 August	REDDITCH
6-7 August	LLANDUDNO
8 August	ABERYSTWYTH
9 August	LLANELLY
10-11 August	PONTYPOOL
12 August	CHELTENHAM
13 August	WANTAGE
14 August	GLASTONBURY
15 August	GILLINGHAM [Dorset]
16 August	CREWKERNE
17 August	BRISTOL
18 August	CRICKLADE
19 August	TAUNTON
20 August	WOOLACOMBE
21 August	TAVISTOCK
22 August	Boskenso Farm, FALMOUTH
23 August	Bolt Head, SALCOMBE
24 August	Salcombe Hill, on A3052, SIDMOUTH
25 August	Sussex farm, West Clandon, GUILDFORD
26 August	Burnt Mills Rd, Nevendon, PITSEA
27 August	Marsh Lands, Minster, SHEERNESS
28 August	Littlestone Aerodrome, HYTHE
29 August	Salts farm, Romney Rd, RYE
30 August	Stud Farm?, St Peters Road, SEAFORD
31 Aug-1 Sept	Sanderstead Park Estate, Limpsfield Rd, SANDERSTEAD
2 September	Wood Oak Farm, RICKMANSWORTH
3 September	Lynn Rd, Walsoken, WISBECH
4 September	Louth Rd, HORNCASTLE
5 September	SUTTON COLDFIELD
6 September	LYTHAM ST ANNES
7 September	Squires Gate Aerodrome, BLACKPOOL
8 September	CANNOCK
9 September	BAKEWELL
10 September	ORMSKIRK
11 September	ALTRINCHAM
12 September	LLANGEFNI
13 September	TUNSTALL
14-15 Sept	Middleton Park, LEEDS
16 September	Wilsthorpe Lane, LONG EATON
17 September	Boardsides, Sleaford Rd, BOSTON
18 September	Rushley Farm, Derby Rd, MANSFIELD
19 September	Waltham Aerodrome, GRIMSBY
20 September	Bleak House Farm, TADCASTER
21 September	Rowe Green Farm, PRESTON
22 September	Seddons Farm, Ainsworth Rd, BURY
23 September	Wernllwyd Farm, WELSHPOOL
24 September	The Aerodrome, PERSHORE
25 September	Pengham Moors Aerodrome, CARDIFF
26 September	Vennaway Lane, Penard, SWANSEA
27 September	Five Lanes, Newport Rd, CHEPSTOW
28 September	Savernake Aerodrome, Burbage Rd, MARLBOROUGH
29 September	Old Barn, Hildenborough, TONBRIDGE

BRITISH HOSPITALS AIR PAGEANT
[Barker & McEwen King]
Tour of UK – 1933

Date	Location
1-2 April	LUTON
8-9 April	Yeading Lane, HAYES
10 April	SUNDRIDGE
11 April	CHATHAM
12 April	Penshurst Aerodrome, TONBRIDGE
13 April	WATFORD
14-15 April	ILFORD
16 April	Hook Aerodrome?, KINGSTON-UPON-THAMES
17-18 April	WALTHAMSTOW
19 April	ADDINGTON
20 April	HITCHIN
21-22 April	HARROW
23 April	CHINGFORD
24-25 April	none
26 April	HANWORTH
27 April	KETTERING
28 April	OXFORD
29 April	GUILDFORD
30 April	Obelisk Farm, Boughton, NORTHAMPTON
1 May	BUCKINGHAM
2 May	LEAMINGTON
3 May	STRATFORD-ON-AVON
4 May	SHREWSBURY
5 May	LUDLOW
6 May	Brockworth Aerodrome, GLOUCESTER
7 May	Woodley Aerodrome, READING
8 May	WINCHESTER
9 May	none
10 May	SWINDON
11 May	STROUD
12 May	TROWBRIDGE
13 May	High Post Aerodrome, SALISBURY
14 May	FAREHAM
15 May	Oakley Wood, WALLINGFORD
16 May	COLCHESTER
17 May	BISHOPS STORTFORD
18 May	CAMBRIDGE
19 May	PETERBOROUGH
20 May	COVENTRY
21 May	ROSS-ON-WYE
22 May	LLANDRINDROD WELLS
23 May	WELSHPOOL
24 May	STAFFORD
25 May	none
26 May	PONTEFRACT
27 May	YORK
28-29 May	LEEDS
30 May	MELTON MOWBRAY
31 May	LINCOLN
1-2 June	BOSTON
3-4 June	STOKE-ON-TRENT
5-6 June	Parc le Breos Farm, Penmaen, SWANSEA
7 June	YEOVIL
8 June	Haldon Aerodrome,TEIGNMOUTH
9 June	TIVERTON
10-11 June	BIRMINGHAM
12-13 June	none
14 June	HIGH WYCOMBE
15 June	WORCESTER
16-17 June	HAVERFORDWEST
18 June	HEREFORD
19 June	FROME
20 June	CAMELFORD
21 June	PLYMOUTH
22 June	ST AUSTELL
23 June	TRURO
24 June	MINEHEAD
25 June	CIRENCESTER
26 June	SHERBORNE
27 June	BUDE
28 June	TAVISTOCK
29-30 June	TAUNTON
1-2 July	BIDEFORD [Heanton Court, Barnstaple?]
3 July	BLANDFORD
4 July	CLEVEDON
5 July	DEVIZES
6 July	WESTON-SUPER-MARE
7 July	DORCHESTER
8 July	PORTSMOUTH
9 July	BOGNOR REGIS
10-11 July	none
12 July	BOURNEMOUTH
13 July	none
14 July	FORD
15-16 July	Shoreham Aerodrome, BRIGHTON
17 July	Rookery Farm, LOWER KINGSWOOD
18 July	WYE
19 July	EASTBOURNE
20 July	CHICHESTER
21 July	Bekesbourne Aerodrome, CANTERBURY
22 July	WHITSTABLE
23 July	DAGENHAM
24 July	CLACTON
25 July	Eldo House Farm, BURY ST EDMUNDS
26-27 July	SOUTHEND
28 July	Holders Farm, Springfield, CHELMSFORD
29 July	none

30 July	GREAT YARMOUTH
31 July	FELIXSTOWE
1 August	Lodge Farm, THETFORD
2 August	KINGS LYNN
3 August	SPALDING
4 August	LONG EATON
5 August	DERBY
6 August	LIVERPOOL
7 August	SPEKE
8 August	SOUTHPORT
9 August	LANCASTER
10 August	CARLISLE
11 August	HOUGHTON-LE-SPRING
12 August	NEWCASTLE
13 August	SHERBURN-IN-ELMET
14 August	NORTHALLERTON
15 August	STOCKPORT
16 August	NUNEATON
17 August	NEWARK
18 August	RUGBY
19 August	WALSALL
20 August	BIRMINGHAM
21 August	LEOMINSTER
22 August	BANBURY
23 August	KIDDERMINSTER
24 August	Ashby, SCUNTHORPE
25 August	STRETFORD
26 August	SHEFFIELD
27 August	HARROGATE
28 August	REDCAR
29 August	WEST HARTLEPOOL
30 August	EATON
31 August	STOCKTON
1 September	CASTLE DOUGLAS
2 September	AYR
3 September	DUNDEE
4 September	ARBROATH
5 September	BANFF
6 September	ABERDEEN
7 September	HUNTLY
8 September	STONEHAVEN
9 September	ST ANDREWS
10 September	RENFREW
11 September	FALKIRK
12-14 Sept	none
15 September	JEDBURGH
16 September	PEEBLES
17 September	DUNBAR
18 September	EDINBURGH
19 September	TYNEMOUTH
20 September	none
21 September	BISHOP AUCKLAND
22 September	DARLINGTON
23 September	YEADON
24 September	none
25 September	DONCASTER
26-27 Sept	none
28 September	HINCKLEY
29 September	WOLVERTON
30 September	SLOUGH
1 October	ABRIDGE
2-5 October	none
6 October	WOKING
7 October	READING
8 October	WOOLWICH

SKY DEVILS AIR CIRCUS TOUR
[Barker & McEwen King]
Tour of UK – 1934
Very few details are known of this tour although 180 towns are reported to have been visited.

15 April	Stag Lane Aerodrome, EDGWARE
24 April	Rookery Farm, LOWER KINGSWOOD
27 April	Oakley Wood, BENSON
29 April	Penshurst Aerodrome, TONBRIDGE
2 May	Fachell Farm, KINMEL BAY
3 June	Wenvoe Aerodrome, CARDIFF
17 August	Cogan Hall Farm, PENARTH
18 August	Whitchurch Aerodrome, BRISTOL
24 August	KINGSNORTH
25 September	Rock Hill Farm, CHIPPING NORTON
30 September	Claybury, WOODFORD [Essex]

FLYING FAIR
[Ronald Dixie Gerran/Aviation Developments Ltd]
UK Tour – 1934
Few details are known of this tour.

31 Mar-2 Apr	Rochford Aerodrome, SOUTHEND
4-5 April	ROMFORD
7-9 April	WATFORD
10 April	HERTFORD
12 April	STAINES
13-14 April	ST ALBANS
15-16 April	AYLESBURY
18 April	HUNTINGDON
21 April	LITTLE WELDON

JUBILEE AIR DISPLAYS
[Barker & McEwan King]
Tour of UK – 1935
Known details of this tour:

25 April	Alderstead Heath, MERSTHAM
28 April	Penshurst Aerodrome, TONBRIDGE
5 May	KIDDERMINSTER
6 May	BIRMINGHAM
7 May	WOLVERHAMPTON
8 June	Vennaway Lane, Parkmill, SWANSEA
9 June	Lock's Common, PORTHCAWL
10-11 June	Wenvoe Aerodrome, CARDIFF
16 August	Racecourse Farm, BODMIN
17 August	Ventomwyn Farm, ST AUSTELL [cancelled due to crash 16.8.35
14 September	COCKERMOUTH
15 September	BLACKBURN

Other known locations include: Walsall, Castle Douglas, Chester, Dartford, Renfrew, Guildford, Harrow, Ireland [tour], Lincoln, Newmarket, Peebles & Pitlochry.

CWA SCOTT'S FLYING DISPLAYS LTD
Tour of the UK – 1936

8 April	Ace of Spades Aerodrome, Hook, KINGSTON-UPON-THAMES
9 April	CHESHAM
10 April	Phoenice Farm, Bagden, DORKING
11 April	Aerodrome, REDHILL
xx April	BARNET
18 April	SLOUGH
19 April	WITLEY
20 April	BURGESS HILL
21 April	BILLINGSHURST
22 April	SWINDON
23 April	DORCHESTER
24 April	TAUNTON
25 April	TORRINGTON
26 April	EXETER
27 April	FROME
28 April	SHEPTON MALLET
29 April	BRISTOL
30 April	MALVERN
8 May	LLANGEFNI
10 May	Phoenix Park, DUBLIN
11 May	Raheny, DUBLIN
12 May	MARYBOROUGH
13 May	Clooncoose Racecourse, LONGFORD
14 May	Big Meadow, ATHLONE
15 May	THURLES
16 May	COBH
17 May	Farmers Cross, CORK
18 May	DUNGARVEN
19 May	MALLOW
20 May	ex-RAF Aerodrome, FERMOY
21 May	CLONMELL
22 May	The Racecourse, TRALEE
23 May	NEWCASTLE WEST
24 May	Banemore, Ballycummin, LIMERICK
25 May	Dunmore, KILKENNY
26 May	Arklow, TEMPLERANEY
27 May	Coolpeach, Drinagh, WEXFORD
28 May	Kilcowen Park Racecourse, WATERFORD
29 May	Silver Strand, Kilpoole, WICKLOW
30 May	Marsh Farm, Dublin Road, DUNDALK
31 May	Church Field, Mornington, DROGHEDA
1 June	Townland, LURGAN
2 June	Upper Broughshane, BALLYMENA
3 June	Ards Aerodrome, BELFAST
4 June	Old Lodge Road, COLERAINE
5 June	Coolkeeragh, DERRY
6 June	ex-RAF Aerodrome, OMAGH

7 June	Scardenmore, Strandhill Road, SLIGO
8 June	Boyle, BALLMORE
9 June	ex-RAF Aerodrome, CASTLEBAR
10 June	Tuam Racecourse, BALLINROBE
11 June	Oranmore Aerodrome, GALWAY
12 June	Castleclare, ENNIS
13 June	Tuam Racecourse, BALLINROBE
14 June	Knockalton Lane, NENAGH
15 June	Naas, LEWISTOWN
16 June	Hollyfort House, ENNISCARTHY
17 June	Old Racecourse, CARLOW
18 June	Leopardstown Racecourse, DUBLIN
19 June	LLANFAIRFECHAN
22 July	SOUTHEND
25 July	Nonsuch Park, Cheam, SUTTON
19-20 August	The Aerodrome, WESTON-SUPER-MARE
5 September	RAMSGATE [cancelled]

Other known locations included: Pitsea, Southwold

BRITISH EMPIRE AIR DISPLAY
[Barker & McEwan King]
Tour of the UK – 1936

8 April	LUTON
9 April	BEDFORD
10 April	OXFORD
11 April	SYWELL
12 April	PETERBOROUGH
13 April	SHEFFIELD
14 April	DONCASTER
15-16 April	none
17 April	NEWARK
18 April	LEAMINGTON SPA
19 April	WOLVERHAMPTON
20 April	RUNCORN
21 April	HEYWOOD
22 April	STAFFORD
23 April	SHREWSBURY
24 April	KIDDERMINSTER
25 April	WALSALL
26 April	NEWCASTLE-UNDER-LYME
27 April	STONE
28 April	CHESTER
29 April	WREXHAM
30 April	WARRINGTON
1 May	Oldfield, Kings Acre, HEREFORD
2 May	none
3 May	none
4 May	REDDITCH
5 May	ROSS-ON-WYE
6 May	none
7 May	WINCHESTER
8 May	MAIDENHEAD
9 May	BASINGSTOKE
10 May	Westwood Heath, COVENTRY
19 May	Bekesbourne Aerodrome, CANTERBURY
20 May	Penshurst Aerodrome, TONBRIDGE
25 May	Whitchurch Aerodrome, BRISTOL
4 June	NEWCASTLE-UPON-TYNE
5 June	LANARK
6 June	CARLISLE
7 June	none
8 June	STRANRAER
9 June	PENICUIK
10-12 June	none
13 June	MACMERRY
14 June	RENFREW
15 June	COCKERMOUTH
16 June	none
17 June	RIPON
18 June	East Leys Farm, BRIDLINGTON
19 June	SKEGNESS
20 June	GAINSBOROUGH
21 June	LEEDS
22 June	HUNTINGDON
23 June	SHEERNESS
24 June	TONBRIDGE
25 June	GRAVESEND
26 June	PULBOROUGH
27 June	MAIDSTONE
28 June	The Aerodrome, ABRIDGE
29 June	Maysland Farm, Great Easton, GREAT DUNMOW
30 June	ROYSTON
1 July	Galleywood Rd, Great Baddow, CHELMSFORD
2 July	CAMBRIDGE
3 July	BURTON-ON-TRENT
4 July	Clifton Aerodrome, YORK

5 July	DEWSBURY
6 July	none
7 July	FARINGDON
8 July	SWINDON
9 July	MINEHEAD
10 July	BRIDGWATER
11 July	SALISBURY
12 July	EASTBOURNE
13 July	BOGNOR REGIS
14 July	WORTHING
15 July	FAREHAM
16 July	UCKFIELD
17 July	LINGFIELD
18 July	none
19 July	DARTFORD
20 July	Rookery Farm, LOWER KINGSWOOD
21 July	STRATFORD-ON-AVON
22 July	The Vale, Craycombe, EVESHAM
23 July	none
24 July	AMMANFORD
25 July	Pengam Moors Aerodrome, CARDIFF
26 July	Lock's Common, PORTHCAWL
27 July	ABERYSTWYTH
28 July	HOLYHEAD
29 July	CAERNARFON
30 July	LLANDUDNO
31 July	LOUGHBOROUGH
1 August	GOOLE
2 August	SHERBURN-IN-ELMET
3-4 August	GREAT YARMOUTH
5 August	IPSWICH
6 August	ST ALBANS
7 August	DEAL
8 August	none
9 August	The Aerodrome, RAMSGATE
10 August	RYE
11 August	BOGNOR REGIS
12 August	none
13 August	NEWBURY
14 August	DERBY
15 August	STOKE-ON-TRENT
16 August	Haydock Park, ASHTON-IN-MAKERFIELD
17 August	SOUTHPORT
18 August	BARROW
19 August	none
20 August	HAWICK
21 August	KELSO
22 August	EDINBURGH
23 August	RENFREW
24 August	FALKIRK
25 August	STIRLING
26 August	KIRKCALDY
27 August	ANSTRUTHER
28 August	ST ANDREWS
29 August	PERTH
30 August	Dyce Aerodrome, ABERDEEN
31 August	FRASERBURGH
1 September	LOSSIEMOUTH
2-4 Sept	none
5 September	SHEFFIELD
6 September	BIRMINGHAM

CORONATION AIR DISPLAYS
[Barker & McEwan King/Aircraft Demonstrations Ltd]
Tour of UK – 1937
Some forty towns were visited.

23 April	Galleywood Rd, Great Baddow, CHELMSFORD
5 May	Aerodrome, REDHILL
6 May	Bekesbourne Aerodrome, CANTERBURY
30 May	Phoenix Park, DUBLIN
31 May	DROGHEDA
1 June	LONGFORD
2 June	PORTLAOGHISE
3 June	CARLOW
4 June	CLONMEL
5-6 June	CORK
7 June	MITCHELSTOWN
8 June	KILKENNY
9 June	ENNISCORTHY
10 June	WATERFORD
11 June	THURLES
12 June	NEWCASTLE WEST
13-14 June	LIMERICK
10 July	CORK

Other locations visited included:
Ramsgate, Sheffield, Greystones, Bundorran.

BIBLIOGRAPHY

The following published books contain further information relating to the subject matter or have been used as reference sources in this book:

Balfour Christopher	*Spithead Express – The Pre-War Island Air Ferry*	Magna Press 1999
Barnes CH	*Handley Page Aircraft Since 1907*	Putnam 1976
Bramson Alan & Birch Neville	*The Tiger Moth Story*	Cassell 1964
Brooks-Pazmany KL	*United States Women in Aviation 1919-1929*	Smithsonian Institution Press 1983
Bruce Hon Mrs Victor	*Nine Lives Plus – Record Breaking on Land, Sea & in the Air*	Pelham Books 1977
Caidin Martin	*Barnstorming – The Great Years of Stunt Flying*	Duell, Sloan & Pearce 1965
Chapman Ted	*Cornwall Aviation Company*	Glasney Press 1979
Cobham Sir Alan	*A Time to Fly*	Shepheard-Walwyn 1978
Cooper AL	*On the Wing*	Blackhawk Publishing 1993
Cruddas Colin	*In Cobhams' Company*	Cobham plc 1994
Dwiggins Don	*Flying Daredevils of the Roaring Twenties*	Arthur Barker 1969
Fahie Michael	*A Harvest of Memories – The Life of Pauline Gower MBE*	GMS Enterprises 1995
Golin Alfred	*No Longer an Island – Britain & the Wright Brothers 1902-09*	Heinemann 1984
Grace Dick	*Crash Pilot*	Longmans Green 1956
Jackson AJ	*Avro Aircraft Since 1908* [2nd edition]	Putnam Aeronautical Books 1990
	De Havilland Aircraft Since 1909 [3rd edition]	Putnam Aeronautical Books 1987
Komons NA	*Bonfires to Beacons*	Smithsonian Institution Press 1989
Lincke Jack R	*Jenny Was No Lady – The Story of the JN-4D*	WW Norton 1970
Longyard William H	*Who's Who in Aviation History*	Airlife 1994
Marrero F	*Lincoln Beachey – The Man who Owned the Sky*	Scottwall Associates 1997
Masefield Sir Peter	*Surrey Aeronautics & Aviation 1785-1985*	Phillimore & Co 1993
Middleton Don H	*Airspeed – The Company and its Aeroplanes*	Terence Dalton 1982
Oliver David	*Hendon Aerodrome – A History*	Airlife 1994
O'Neil Paul	*Barnstormers & Speed Kings*	Time-Life Books 1981
O'Rourke Madeleine	*Air Spectaculars – Air Displays in Ireland*	Glendale 1989
Ronnie Art	*Locklear: The Man who Walked on Wings*	AS Barnes & Co 1973
Smith Elinor	*Aviatrix*	Harcourt Brace Jovanovich 1981
Smith Vi	*From Jennies to Jets – The Aviation History of Orange County*	Sultana Press 1974
Tessendorf KC	*Barnstormers and Daredevils*	Macmillan Publishing 1988
Thorp Stan	*Aeronautical Memoirs*	Brian Thorp 1988
Ward H & Hearn P	*The Yorkshire Birdman – Memoirs of a Pioneer Parachutist*	Robert Hale 1990
Waterman WD	*Waldo – Pioneer Aviator 1910-1944*	Arsdalen Bosch 1988
Wynne H Hugh	*The Motion Picture Stunt Pilots*	Pictorial Histories 1987

INDEX
TO PERSONALITIES AND OPERATORS

AIR-BRITAIN SALES

This publication and companion volumes are available by post-free mail order from

Air-Britain Sales Department (Dept TFFY)
41 Penshurst Road, Leigh,
Tonbridge, Kent TN11 8HL

For a full list of current titles and details of how to order, visit our e-commerce site at www.air-britain.co.uk
Visa / Mastercard / Delta / Switch accepted - please give full details of card number and expiry date.

THE TRIPLE ALLIANCE - The Predecessors of the first British Airways
By Neville Doyle   £18.00 (Members) £22.50 (Non-members)
The story of Hillman's Airways Ltd, Spartan Air Lines and United Airways Ltd, and other operators involved through mergers and take-overs, who amalgamated on 1st October 1935 to form British Airways Ltd. Fully illustrated with contemporary photos, timetables and advertising, and describing the personalities, routes flown and aircraft used. Hardback A4, 128 pages with 8 in colour.

THE de HAVILLAND DRAGON/RAPIDE FAMILY
By John F Hamlin   *Available August 2003* £19.95 (Members) £29.95 (Non-members)
The story of the development of the de Havilland Dragon series of biplane airliners which covers the complete histories of all individual DH.84, DH.86, DH.89 and DH.90 models produced. Fully illustrated including16 pages of colour photos and drawings. Hardback A4, 256 pages.

THE SOPWITH PUP
By J M Bruce, Gordon Page and Ray Sturtivant   £24.00 (Members) £30.00 (Non-members)
The definite work on the Sopwith Pup, covering in detail the aircraft's development, service history and units in a narrative account supported by all known details of individual aircraft histories, survivors and replicas. Hardback A4, 320 pages containing over 400 photographs and numerous colour and black-and-white scale drawings.

THE DH.4/DH.9 FILE
By Ray Sturtivant and Gordon Page   £24.00 (Members) £30.00 (Non-members)
The story of the development of these types built in large numbers for the RFS, RNAS, RAF and USAAC. Another definitive work which includes details of widespread use overseas. Hardback A4, 300 pages including 450 photos and 16 pages of colour drawings.

SPITFIRE INTERNATIONAL
By Helmut Terbeck, Harry van der Meer and Ray Sturtivant   £32.50 (Members) £39.50 (Non-members)
Well over 6000 Spitfires served with overseas air and naval forces during and after the Second World War. All known details of these, together with backgound information about their operators and the units to which they belonged, and of the Spitfires known to have flown outside the UK with civil registrations, are included in this 480 page A4 hardback book which contains many colour photos, drawings, maps and insignia.

THE DH.106 COMET An Illustrated History
By Martin Painter   £29.50 (Members) £37.00 (Non-members)
The story of the development on Britain's pioneering jet airliner, its operators, routes, detailed individual histories, fates and survivors. Contains over 600 illustrations, 242 in colour. Hardback A4, 368 pages.

For civil aircraft enthusiasts Air-Britain publishes annually a series of titles covering UK and European Registers, worldwide Airline Fleets, Business Jets and a series of low-priced "Quick Reference" books on these topics.
Air-Britain also publishes a comprehensive range of military titles, please check for latest details of RAF Serial Registers, detailed RAF aircraft type "Files", Squadron Histories and Royal Navy Aircraft Histories.

IMPORTANT NOTE - Members receive substantial discounts on prices of all the above Air-Britain publications.
For details of membership see the following page or visit our website at http://www.air-britain.co.uk

AIR-BRITAIN MEMBERSHIP

Join on-line at www.air-britain.co.uk

If you are not currently a member of Air-Britain, the publishers of this book, you may be interested in what we have on offer to provide for your interest in aviation.

About Air-Britain

Formed over 50 years ago, we are the world's most progressive aviation society, and exist to bring together aviation enthusiasts with every type of interest. Our members include aircraft historians, aviation writers, spotters and pilots – and those who just have a fascination with aircraft and aviation. Air-Britain is a non-profit organisation, which is independently audited, and any financial surpluses are used to provide services to the ever-growing membership. In the last 7 years, our membership has increased annually, and our current membership now stands at over 4,200.

Membership of Air-Britain

Membership is open to all. A basic membership fee is charged and every member receives a copy of the quarterly house magazine, Air-Britain Aviation World, and is entitled to use all the Air-Britain specialist services and to buy **Air-Britain publications at discounted prices**. A membership subscription includes the choice to add any or all of our other 3 magazines, News &/or Archive &/or Aeromilitaria. Air-Britain publishes 10-20 books per annum (around 70 titles in stock at any one time). Membership runs January - December each year, but new members have a choice of options periods to get their initial subscription started.

Air-Britain Aviation World is the quarterly 48-page house magazine containing not only news of Air-Britain activities, but also a wealth of features, often illustrated in colour, on many different aviation subjects, contemporary and historical, contributed by our 4,200 members.

Air-Britain News is the world aviation news monthly, containing data on Aircraft Registrations worldwide, and news of Airlines and Airliners, Business Jets, Local Airfield News, Civil and Military Air Show Reports and International Military Aviation. An average 160 pages of lavishly–illustrated information for the dedicated enthusiast.

Air-Britain Archive is the quarterly 40-48 page specialist journal of civil aviation history. Packed with the results of historical research by Air-Britain specialists into aircraft types, overseas registers and previously unpublished photographs and facts about the rich heritage of civil aviation. Up to 100 photographs per issue, some in colour.

Air-Britain Aeromilitaria is the quarterly 48-page unique source for meticulously researched details of military aviation history edited by the acclaimed authors of Air-Britain's military monographs, featuring British, Commonwealth, European and U.S. Military aviation articles. Illustrated in colour and black & white.

Other Benefits

Additional to the above, members have exclusive access to the Air-Britain e-mail Information Exchange Service (ab-ix) where they can exchange information and solve each other's queries, and to an on-line UK airfield residents database. Other benefits include numerous Branches, use of the Specialists' Information Service; Air-Britain trips and access to black & white and colour photograph libraries. During the summer we also host our own popular FLY-IN. Each autumn, we host an Aircraft Recognition Contest.

Membership Subscription Rates – from £10 per annum.

Membership subscription rates start from as little as £10 per annum, and this amount provides a copy of 'Air-Britain Aviation World' quarterly as well as all the other benefits covered above. Subscriptions to include any or all of our other three magazines vary between £18 and £50 per annum (slightly higher to overseas).

Join on-line at www.air-britain.co.uk or, write to 'Air-Britain' at 1 Rose Cottages, 179 Penn Road, Hazlemere, High Wycombe, Bucks HP15 7NE, UK, or telephone/fax on 01394 450767 (+44 1394 450767) and ask for a membership pack containing the full details of subscription rates, samples of our magazines and a book list.